COOL ENERGY

THE RENEWABLE SOLUTION TO GLOBAL WARMING

Michael Brower

A REPORT BY THE
UNION OF CONCERNED SCIENTISTS

Table of Contents

List of Figures v

List of Tables vii

Foreword ix

Acknowledgments xi

A Note on Units of Measure xiii

Introduction 1

Chapter one
Fossil Fuels and Global Warming 5
    The Global Climate Experiment 5
    The Role of Fossil Fuels 8

Chapter two
Energy Strategies 11
    Energy Efficiency 12
    Nuclear Power 14
    Renewable Energy Sources 18
    Mitigating Global Warming 24

Chapter three
Solar Energy 27
    The Solar Resource 28
    Technologies and Economics 28
    Energy Storage 41
    Conclusions and Near-Term Prospects 42

Chapter four
Wind Energy **45**
    The Wind Resource **46**
    Technologies and Economics **46**
    Energy Storage **51**
    Environmental Impacts **51**
    Conclusions and Near-Term Prospects **52**

Chapter five
Biomass Energy **53**
    The Biomass Resource **54**
    Technologies and Economics **57**
    Environmental Impacts **63**
    Conclusions and Near-Term Prospects **65**

Chapter six
Hydroelectric Power **67**
    The Hydroelectric Resource **67**
    Technologies and Economics **69**
    Environmental Impacts **70**
    Conclusions and Near-Term Prospects **71**

Chapter seven
The Path to a Renewable Future **73**
    Market Barriers **73**
    Policy Recommendations **78**

References **81**

List of Figures

Figure 1 Atmospheric CO_2 Concentration **7**

Figure 2 Projections of Global Warming **7**

Figure 3 US CO_2 Emissions by Sector **8**

Figure 4 World Carbon Emissions
from Fossil Fuels **9**

Figure 5 Energy Consumption and GNP **13**

Figure 6 Global Energy Use in 2020 ....**13**

Figure 7 Capital Costs of Nuclear Power **16**

Figure 8 Costs of Electricity **21**

Figure 9 Costs of Ethanol and Gasoline **21**

Figure 10 US Energy Supply
in 1989 and 2020 ....**23**

Figure 11 Sales of Solar Collectors **31**

Figure 12 A Photovoltaic Cell **36**

Figure 13 Photovoltaic-Dish System Cost vs.
Annual Production **40**

Figure 14 California Wind-Power Plants **47**

Figure 15 Wind Resource Map **48**

Figure 16 Components of a Wind Turbine **50**

Figure 17 Biomass Energy
Consumption in 1987 and 2000 ....**58**

Figure 18 Pathways for Biofuel Production **61**

Figure 19 Oil-Price Projections **75**

Figure 20 Renewable-Energy
R&D Funding **77**

List of Tables

Table 1 Carbon-Dioxide Emissions from Fossil Fuels 9

Table 2 Renewable Energy Resources 19

Table 3 UCS Scenario of Renewable-Energy Supply in 2000 and 2020 22

Table 4 Photovoltaic-Cell Efficiencies 38

Table 5 Biomass Resource Estimates 55

Table 6 Comparison of Biomass Resource Estimates with Other Studies 55

Foreword

The prospect of global warming from the greenhouse effect is now attracting intense public concern. Growing interest is being devoted to averting and coping with the disastrous worldwide consequences that are likely if vigorous action is not taken soon. Few people are aware of the promise of relatively benign, renewable energy sources—solar, in its many forms, wind, biomass, and others—to augment and eventually replace fossil fuels, the source of most greenhouse emissions. These technologies have been steadily and quietly improving over the years, in spite of the government's policy of neglect. This careful study by Dr. Michael Brower comes at a key moment when public concern may soon be translated into government action. It should make clear to doubters and to the uninformed that a renewable future is within our reach. A sound national program to develop and implement alternative energy sources would greatly lessen the threat of global warming. COOL ENERGY is an excellent guide for those who recognize a responsibility to meet this grave challenge to humanity's future.

Henry W. Kendall
Chairman of the Board
Union of Concerned Scientists

Acknowledgments

Many people provided invaluable help to me in researching and writing this report. I am particularly indebted to the reviewers: Eldon Boes (Sandia National Laboratory), Stanley Charren (Kenetech Corporation), Tom Gray (American Wind Energy Association), Patricia Layton (Oak Ridge National Laboratory), Lee Lynd (Dartmouth College), James MacKenzie (World Resources Institute), Larry Mansueti (American Public Power Association), B.W. Marshall (Sandia National Laboratory), Kevin Porter (Investor Responsibility Research Center), David Rinebolt (National Wood Energy Association), Scott Sklar (Solar Energy Industries Association), Blair Swezey (Solar Energy Research Institute), Anthony Turhollow (Oak Ridge National Laboratory), and Susan Williams (Investor Responsibility Research Center). I am also grateful to Steven Krauss for his editorial skill and Herb Rich for the design and production of the report.

M.C.B.

Cover photos (l to r): Luz International, U.S. Windpower, Department of Energy

A Note on Units of Measure

This report uses metric units. The base unit of energy consumption is thus the joule; one joule is approximately the energy required to raise a kilogram weight 10 centimeters. Primary-energy consumption—before conversion to electricity—is expressed in exajoules (EJ), where one exajoule equals 10^{18} joules; the United States consumes about 84 exajoules annually. The cost of primary energy is expressed in dollars per gigajoule (GJ), where a gigajoule is 10^9 joules; a gasoline price of $1 per gallon is equivalent to an energy cost of about $9 per gigajoule. For those more familiar with English units, a British thermal unit, or Btu, is equal to 1055 joules; thus, an exajoule is almost exactly equal to a quadrillion Btu, or quad, and a gigajoule is almost exactly equal to a million Btu.

The base unit of electrical power is the watt (joule per second). Electrical generating capacity is expressed in kilowatts (kW) or megawatts (MW), and the quantity of electricity produced is expressed in kilowatt-hours (kWh). One kilowatt-hour equals 3.6 megajoules (MJ); because of conversion losses, however, about 10.9 megajoules of primary energy are actually needed to generate a kilowatt-hour of electricity in a conventional steam-electric plant. The latter figure is used to calculate the quantity of fossil fuels displaced by renewable and nuclear sources. For example, wind turbines generate approximately 1.8 billion kilowatt-hours of electricity annually, but they are said to displace the primary-energy equivalent of 0.02 exajoules.

Introduction

"Our civilization," wrote George Orwell in *The Road to Wigan Pier* (1937), "is founded on coal." Orwell's observation, updated to reflect the advent of petroleum and natural gas, is as true today as it was 50 years ago. Fossil fuels heat our homes and generate our electricity, run our cars and power our industries. Without them, it is safe to say, the United States and other industrialized countries would not have achieved the great prosperity and power they now enjoy and to which less-developed countries aspire.

But the world cannot continue to rely so heavily on fossil fuels without placing the global environment at risk. Acid rain and air pollution produced by automobiles, electric-power plants, and other fossil-fuel sources are damaging trees, crops, and human health on a regional and global scale. Even more important, the earth could soon experience unprecedented temperature increases and disruptive climate changes brought about by the accumulation of atmospheric "greenhouse" gases. These heat-absorbing gases are released by a variety of human activities, chief among them the burning of fossil fuels. If present trends in greenhouse-gas emissions continue, the earth could be committed to a warming of 1.5–4.5 degrees Centigrade ($^{\circ}$C) by the middle of the next century, with potentially disastrous consequences for agriculture, forests, wildlife, coastal communities, and the quality of human life.

Although it is too late to prevent some warming from occurring, its magnitude can be greatly diminished if action is taken soon. The key is to reduce world consumption of fossil fuels by 50 percent or more over the next several decades. But can deep cuts be accomplished without sacrificing economic growth and prosperity? Conventional wisdom says no; fossil fuels are simply too cheap and convenient to do without. Yet a closer examination of the alternatives suggests a different answer. Deep cuts in fossil-fuel use are indeed possible and may even be necessary to ensure long-term prosperity. They cannot be achieved, however, without a radical shift in energy policies, particularly by the United States, the world's leading energy consumer.

One crucial step toward reducing the threat of global warming will be to raise the efficiency of energy use—cutting, in one stroke, consumption of fossil fuels and the release of pollutants and greenhouse gases. Studies have shown that the United States and other industrialized countries could reduce their per-capita energy consumption by as much as one-half by adopting more efficient technologies in transportation, building construction, industry, and other economic sectors. Far from causing economic hardship, making more efficient use of energy could help our economy by reducing energy expenditures and limiting oil imports.

Energy efficiency alone will not be enough, however. To minimize the risk of global warming, fossil fuels will have to be replaced by energy sources that do not emit greenhouse gases. The most likely alternatives are renewable energy sources, drawn from vast and inexhaustible resources of sunlight, winds, oceans, rivers, and plants. Once considered exotic and impractical, the technologies for exploiting these resources are becoming increasingly reliable and cost effective in comparison to conventional energy technologies. Some are already widely successful and today supply almost 8 percent of US energy demand. Others—particularly wind and solar technologies, as well as processes that convert biomass (plant matter) to liquid and gaseous fuels—are now or soon could be competitive with fossil fuels in a broad range of applications. Although some technical issues remain to be solved, there appear to be no insurmountable barriers to prevent renewable energy sources from eventually meeting most, if not all, of US and world energy needs.

But the promise of renewable energy sources will not be realized without strong government leadership. Amidst the oil "crises" of the 1970s, a great deal of attention was given to developing renewable energy sources as a way to reduce oil imports. However, interest has waned since then as oil prices have fallen and supplies have become more plentiful. The Reagan administration and the Congress shortsightedly cut funding for renewable-energy research and development by more than 80 percent from 1981 to 1989 and eliminated tax credits for most renewable-energy investments. As a result, industry growth has slowed and in many cases reversed, and US manufacturers of renewable-energy systems have been losing ground to foreign competitors—many of whom enjoy better support from their governments—in a pattern reminiscent of the decline of the domestic consumer-electronics industry in the 1970s.

As public concern about the threat of global warming has risen, government attention has turned not to developing renewable energy sources but to the possibility of reviving nuclear power. Yet although nuclear power could help reduce fossil-fuel use and global warming over the long run, its near-term role is likely to be limited. The present generation of nuclear-reactor designs is at an impasse, having lost the confidence of the public and electric utilities because of rising costs and accidents like that at Three Mile Island. The development and demonstration of new and possibly safer reactors will take several years. Even if new designs prove acceptable, other troublesome issues will remain, in particular the safe disposal of radioactive wastes and the prevention of nuclear-weapons proliferation.

Renewable-energy technologies raise none of these problems. Most are nonpolluting and pose little risk to public safety. Their technical characteristics and performance, in general, are well understood. And unlike nuclear power, which generates only electricity, they can supply directly all of the principal forms of energy we use today: low- and high-temperature heat, electricity, and transportation fuels. The one major drawback of solar and wind power is that they are inherently variable and so require costly energy storage to be completely reliable. For many near-term applications, however, storage will not be needed as long as solar and wind power are a relatively small part of the overall energy-supply mix, and in those instances where it is necessary, fossil fuels—particularly natural gas—could serve as an acceptable interim backup while more cost-effective storage technologies are developed.

The development of renewable energy sources would bring not only environmental benefits but economic ones as well. It would help reduce oil imports, which have increased in recent years and may surpass 50 percent of domestic oil consumption in the early 1990s. It would also help insulate our economy against fossil-fuel prices that are almost certain to rise as oil and natural-gas reserves are depleted. And it would create domestic jobs—far more, in all likelihood, than would be lost due to cutbacks in fossil-fuel industries.

The United States must renew its commitment to developing renewable energy sources. We urge, at a minimum, that the federal government adopt the following policies:

- Reinstitute renewable-energy tax credits;
- Increase funding for renewable-energy research and development;
- Modify electric-utility regulations to give greater preference to environmentally benign technologies;
- Buy renewable-energy systems for government facilities; and
- Increase support for renewable-energy exports.

These steps, described more fully later in this report, would cost the government no more than about $10 billion a year by 2000, mainly in reduced tax revenues, and could be paid for by a modest increase in taxes on fossil fuels or a reduction in government subsidies for conventional energy technologies. With the implementation of these policies, we estimate that production of renewable energy in the United States could grow to 15 percent of energy demand by 2000, almost double the present fraction. With further technical progress and policy changes in decades to follow, the renewable fraction could rise as high as 50 percent by 2020, putting the United States—and the world—well on the path to a renewable future.

In this report we examine four renewable energy sources—solar, wind, biomass, and hydroelectric—and attempt to assess their potential role in mitigating global warming. We focus on the United States because it is here that the world will look for leadership in the development of new energy technologies; without a commitment by the United States, it is unlikely that many other countries will take major steps to reduce their fossil-fuel consumption. In concentrating on only four energy sources, we have neglected some others, in particular ocean-thermal energy, tidal power, and geothermal energy. This is not meant to suggest that they could not play an important role in our future energy supply; we simply have not done the research necessary to make a fair assessment of their potential contribution.

Fossil Fuels and Global Warming

In April 1815, the Indonesian volcano Mount Tambora erupted, spewing immense quantities of dust, ash, and debris into the air. The following year, inhabitants of New England experienced the coldest summer on record, with snowstorms occurring in June and frosts continuing through August.

Strange as it may seem, the two events were almost certainly connected. Dust deposited in the upper atmosphere by the eruption absorbed sunlight, causing surface temperatures to drop over much of the globe. Temperatures recorded in Europe and North America in 1816 were 1.0–2.5°C lower than normal. From events such as this, scientists now recognize that volcanic eruptions can have a temporary effect on climate.

Yet the more important lesson to be drawn is that seemingly small changes in atmospheric composition can have profound effects on human lives. The cold spell of 1816 destroyed much of the New England corn crop vital to subsistence farmers, causing near starvation in some areas. In Europe, its effects were even more severe: famines and food riots occurred in France, Switzerland, and other countries, and wheat prices doubled (Schneider and Londer 1984, Stommel and Stommel 1979).

The Global Climate Experiment

Now humanity is conducting its own experiment on the atmosphere, one whose ultimate consequences may be no less severe, and far more enduring, than those of the Tambora eruption. Some of the effects of this experiment are already discernible in the United States. Air pollution is affecting the health and comfort of millions of Americans and causing an estimated 5-10 percent loss in agricultural productivity. Acid rain is damaging forests, lakes, and streams in parts of the eastern United States, and its effects are spreading beyond national borders (WRI 1988, MacKenzie and El-Ashry 1988).

More serious still is the threat of global warming. Scientists have long known that certain gases in the atmosphere absorb heat (infrared light) radiating from the earth's surface. Most of these gases, such as carbon dioxide, methane, nitrous oxide, and ozone, exist naturally, and without their warming influence—the greenhouse effect—the earth would be uninhabitable. However, human activities—including the burning of fossil fuels, tropical deforestation, agricultural practices, and the release of chemicals used in refrigerants and other products—are believed to be enhancing the greenhouse effect by adding ever-greater quantities of greenhouse gases to the atmosphere (Bolin *et al.* 1986, Ramanathan 1988, Schneider 1989). The atmospheric concentration of the most important gas, carbon dioxide, has increased 20-30 percent in the past 100 years (see Figure 1).

Whether the accumulating greenhouse gases have already resulted in a measurable warming of the earth is not yet clear because natural variations in temperature tend to obscure small increases that might occur; an increase of about 0.5°C was observed in temperature records of the past 100 years, but this cannot be ascribed unambiguously to an enhancement of the greenhouse effect. Nevertheless, there is a consensus among atmospheric scientists that the earth is very likely to get warmer in the future. If present trends continue, the concentrations of all greenhouse gases could rise to the equivalent of double the preindustrial concentration of carbon dioxide by the middle of the next century. This could commit the earth to a warming of 1.5–4.5°C.[1] The actual warming would be delayed on the order of decades by the thermal inertia of the oceans (see Figure 2).

Such a sustained and rapid temperature increase would be unprecedented in human history. To put it in perspective, today's global average temperature is only about 5°C warmer than it was at the peak of the last Ice Age 18,000 years ago, when much of the Northern Hemisphere was covered by ice sheets kilometers thick (Schneider 1989). The effects of an equally large—and far more rapid—future warming would be profound and irreversible, and almost certainly adverse to the human race.

A 1988 Environmental Protection Agency study attempted to predict some of the possible effects on the United States of global warming in the next century. The study suggested that forests would begin to decline within a few decades and some species of plants and animals would die out because they could not migrate north quickly enough or their paths of migration would be blocked by urban sprawl. Most of the country's coastal wetlands—many of them irreplaceable wildlife refuges—would be lost to rising seas caused by the melting of land-based polar ice and the thermal expansion of seawater. Coastal communities would have to spend large sums to protect against flooding. Agricultural output would be affected, as forecasters predict increased summer dryness in the American breadbasket and a higher frequency of droughts and heat waves (although increased levels of atmospheric carbon dioxide could partially offset these changes by aiding plant growth). In some parts of the country, water for drinking, irrigation, and industry would become more scarce (EPA 1988).

Atmospheric CO$_2$ Concentration

Figure 1. Historical record of the atmospheric concentration of carbon dioxide, based on direct and indirect measurements. Source: Environmental Protection Agency (1989).

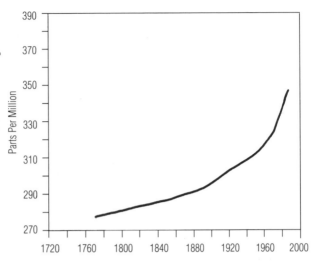

Projections of Global Warming

Figure 2. Projections of global warming based on two scenarios of world economic development, a rapidly changing world (upper range) and a slowly changing world (lower range). Source: Environmental Protection Agency (1989).

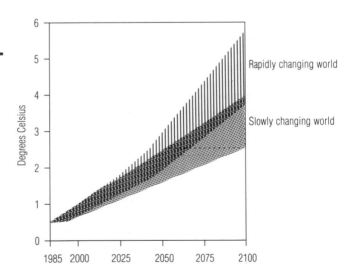

US CO₂ Emissions by Sector, 1988

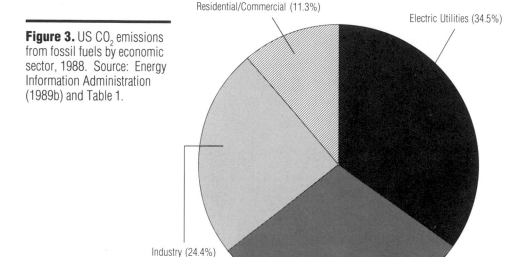

Figure 3. US CO₂ emissions from fossil fuels by economic sector, 1988. Source: Energy Information Administration (1989b) and Table 1.

Residential/Commercial (11.3%)

Electric Utilities (34.5%)

Industry (24.4%)

Transportation (29.8%)

The effects of global warming could well take on tragic proportions in other parts of the world, particularly in less-developed countries ill-equipped to cope with rapidly changing climate conditions. Famines could occur as heavily populated, food-producing coastal regions are inundated by rising seas and the interiors of continents are afflicted by more frequent droughts.

Taken together, the possible effects of global warming present a frightening threat to future generations. Some effects, moreover, may not yet have been anticipated. As the EPA study noted, predictions are "inherently limited by our imaginations...Until a severe event occurs...we fail to recognize the close links between our society, the environment, and climate."

The Role of Fossil Fuels

A variety of human activities are thought to be contributing to global warming by releasing greenhouse gases into the atmosphere. They include the destruction of tropical rainforests and the associated release of carbon dioxide, methane, and other gases; agricultural practices, such as the use of nitrogen-rich fertilizers, which generates nitrous oxide, and the growing of rice in flooded paddies, which produces methane; and emissions of chlorofluorocarbons (CFCs) used in refrigerants, aerosol sprays, foam products, and solvents.

Table 1. CO_2 Emissions Per Unit of Fuel Consumed. Source: Mintzer (1987).

	Kilograms/Gigajoule	Ratio to Natural Gas
Natural Gas	13.8	1.00
Oil	19.7	1.43
Coal	26.9	1.95
Shale Oil	47.6	3.45
Synthetic Oil (from coal)	38.6	2.80
Synthetic Gas (from coal)	40.7	2.95

World Carbon Emissions from Fossil Fuels

Figure 4. World carbon emissions from fossil-fuel use, 1860-1985. Source: Environmental Protection Agency (1989).

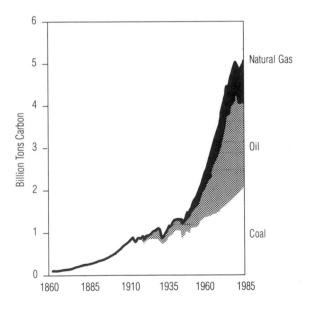

The chief source of greenhouse gases, however, is the combustion of fossil fuels (coal, oil, and natural gas). Worldwide, fossil-fuel combustion accounts for more than 70 percent of all human carbon-dioxide emissions, or approximately 5.6 billion metric tons of carbon each year (Houghton and Woodwell 1989). (Emissions are usually measured by weight of carbon in carbon dioxide.) Fossil fuels are also a source of nitrous oxide, methane, and, indirectly, tropospheric ozone (a component of smog). Nitrous oxide is produced in combustion, methane through leaks from natural-gas wells and pipelines as well as coal mines, and ozone through chemical reactions involving methane, nitrogen oxides, and other compounds. All told, fossil-fuel use accounts for over half of the global warming predicted for the next 100 years (EPA 1989).

Although all countries contribute to global warming, the United States bears an especially heavy responsibility. In 1988, US fossil-fuel use accounted for about 23.7 percent of world carbon-dioxide emissions from fossil fuels. Since 1986, carbon-dioxide emissions have grown proportionately faster in the United States than in the rest of the world, reversing a 17-year trend (Weisskopf 1989). Within the US economy, electric utilities are the largest source of carbon dioxide, emitting about 35 percent of the total, followed by the transportation, industrial, residential, and commercial sectors (see Figure 3). The electric utilities' share is so large in part because 60 percent of electricity is generated from coal. For each unit of energy obtained in combustion, coal emits some 40 percent more carbon dioxide than oil and almost 100 percent more carbon dioxide than natural gas (see Table 1).

World fossil-fuel consumption and carbon-dioxide emissions have almost quadrupled since 1950 (see Figure 4), and without major changes in energy policies they are likely to continue expanding, possibly doubling before 2025 (EPA 1989). Fossil-fuel consumption in the United States is predicted to increase 5-17 percent by 2000 and 13-36 percent by 2010 (DOE 1988a). Even without the threat of global warming, these trends cannot be sustained without causing excessive damage to the environment. As fossil-fuel use grows, it will become increasingly costly and difficult for communities to meet clean-air goals through conventional pollution-control strategies, necessary though they will be. Moreover, such strategies will do little or nothing to reduce emissions of greenhouse gases. A new approach to protecting the global environment is necessary, one that goes to the root of the problem: humanity's addiction to fossil fuels.

Notes

1. The range of temperatures is chosen to account for possible climatic feedback effects, such as changes in cloud cover, as well as differences between climate models. The predicted warming is a global, seasonal average; some regions of the earth, especially those near the poles, could experience greater warming, and the effects in different seasons would vary.

Energy Strategies

A central goal of US energy policy for much of this century has been to ensure reliable and affordable supplies of fossil fuels. At times this has meant taking military action, as when the Navy defended oil shipping against Iranian attacks in the Persian Gulf in 1983–1988. Only rarely, however, has US policy looked toward the future. Even before the threat of global warming attracted attention, it was clear that US and world oil and natural-gas reserves would not last indefinitely, and that prudent planning called for gradually reducing our dependence on these energy sources.

The United States can no longer afford—if it ever could—to be so shortsighted. The threat of global warming and other environmental problems demands a long-term commitment by all nations to making major reductions in fossil-fuel use. The United States, as the chief industrialized nation and largest consumer of fossil fuels, must lead the world toward this goal. Action must be taken soon, as every decade of delay commits the earth to a probable additional warming of 0.2-0.6°C, on top of the 0.5-1.5°C believed already locked in because of previous greenhouse-gas emissions (Mintzer 1987).

In order to stabilize the global climate, according to a recent Environmental Protection Agency study, world carbon-dioxide emissions will have to be cut 50-80 percent from present levels while emissions of other gases will have to be reduced by amounts ranging from 10 percent to 100 percent (EPA 1989). Such large reductions will require a profound restructuring of the world energy economy and therefore cannot be accomplished on a time scale shorter than decades. Nevertheless, the process can be started almost immediately. Participants of the 1988 Toronto World Conference on the Changing Atmosphere recommended, as an initial goal, that carbon-dioxide emissions be reduced 20 percent by 2005, with deeper reductions of 50 percent or more to occur later (Toronto 1988). This timetable, though ambitious, is both technically and economically feasible, in our judgment, and should be adopted by the United States and other countries.

Such a radical change in energy policy is bound to provoke controversy. It is argued in some circles that the scientific uncertainties surrounding global warming are so great as to preclude major policy changes at this stage; further research is all that is called for (see, e.g., Marshall Institute 1989). But although it is possible that the effects of global warming will be less severe than scientists now believe, it is equally likely that they will be more severe. It would be prudent to err on the side of caution, for a mistake in the other direction could have disastrous consequences. Moreover, any reduction in US fossil-fuel use could have other important benefits, such as reducing acid rain and air pollution and limiting oil imports and energy expenditures.

Three principal strategies are available for reducing fossil-fuel use and carbon-dioxide emissions: improving energy efficiency, expanding the use of nuclear power, and developing renewable energy sources.[1] Of these, energy efficiency and nuclear power have received the most attention amidst rising concern about global warming. Renewable energy sources have been less in the public eye, but in some respects they are the most important. To a greater degree than nuclear power, they can provide cost-effective and environmentally benign substitutes for fossil fuels in a broad range of applications. Combined with improvements in energy efficiency—as well as other strategies to reduce emissions of greenhouse gases not produced by fossil fuels—the development of renewable energy sources by the United States and other countries could lead to a virtual halt in global warming toward the end of the next century (EPA 1989).

Energy Efficiency

The past 16 years have seen a revolution in the efficiency of energy use in the United States. Before oil and other energy prices began rising in 1973, energy consumption had increased in direct proportion to economic growth. But since 1973, consumption has stayed almost constant—except for the last three years, during which oil and gas prices have been low—while the economy has expanded almost 45 percent (see Figure 5). The change is due largely to improvements in the efficiency of products ranging from automobiles to air conditioners. Without these improvements, the United States would be spending about $150 billion more for energy each year than it does today and would be importing nearly twice as much oil.

Despite this progress, vast quantities of energy are still being wasted (Chandler *et al.* 1988, Goldemberg *et al.* 1987). The United States uses twice as much energy per dollar of gross national product (GNP) as Japan, West Germany, and some other advanced industrialized nations. The average American house is relatively poorly insulated, requiring 3-10 times as much energy for heating and cooling as energy-efficient or super-insulated houses. And while in 1988 the

Energy Consumption and GNP

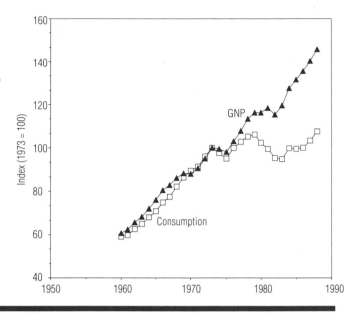

Figure 5. US energy consumption and gross national product, 1960-1988, indexed to 1973. Source: Energy Information Administration (1989b).

Global Energy Use in 2020

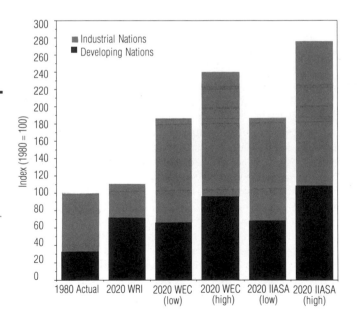

Figure 6. Comparison of the World Resources Institute (WRI) scenario of global primary-energy use in 2020 with low and high projections from two other studies (WEC 1983, IIASA 1981). Adapted from Goldemberg *et al.* (1987).

average highway fuel economy of new cars in the United States was, by official measure, 28.7 miles per gallon (mpg), several foreign and domestic models on the market get over 50 mpg, and prototypes have been developed which exceed 100 mpg (DOT 1989, Bleviss 1988).

By taking advantage of the opportunities for savings, the United States could reduce future energy consumption, or at least limit its growth. According to one study, energy consumption in 2000 could be reduced 7 percent below today's level, or 18 percent below the level projected for that year in official forecasts, if a number of policies are adopted. The policies include raising the national fuel-economy standards on new motor vehicles, adding 50 cents per gallon to the federal gasoline tax, reforming electric-utility regulation to foster investment in end-use energy efficiency, and placing more stringent efficiency standards on electric appliances, lighting, and other products. Far from sacrificing economic growth, these steps would result in estimated annual savings of $75 billion (Geller 1989).

Even greater efficiency gains are possible over the long term. Goldemberg *et al.* (1987) estimate that with current technologies the industrialized countries could cut their per-capita energy use by one-half and total energy use by one-third in 30 years. Over the same period, less-developed countries could raise their standard of living to the level of 1970s Western Europe by adopting more efficient technologies for cooking, lighting, and other basic needs. With these changes, world energy consumption would be only slightly higher in 2020 than it is today and about half of what it is projected to be otherwise (see Figure 6).

But while energy efficiency will be essential to limiting fossil-fuel use, it will not be sufficient by itself to reduce world carbon-dioxide emissions substantially below today's level because of the offsetting effects of population and economic growth. Thus, to minimize the risk of global warming, it will probably be necessary to replace at least 50 percent of fossil-fuel use with alternative energy sources that do not produce greenhouse gases. The major choices are renewable energy sources and nuclear power.

Nuclear Power

During the 1970s, nuclear power was heralded as a promising means of reducing US dependence on imported oil. Indeed, in the past 10 years the production of electricity from nuclear power plants has multiplied sixfold, and as of June 1989 there were 110 operable plants supplying almost 20 percent of the electricity generated in the United States or the equivalent of 6.6 percent of primary-energy consumption (EIA 1989a). Some analysts believe that nuclear power's share could grow much larger in the future, possibly more than doubling before 2025 (Fulkerson, Perry, and Reister 1988).

However, the future of nuclear power in the United States is in serious doubt. No new plants have been ordered since 1978, and all those ordered since 1973 have been canceled. Once the few plants still under construction are completed, no new additions to nuclear capacity are likely for at least the next 10-15 years, as it takes that long to plan, construct, and obtain an operating license for a new plant. Thus, nuclear power can make no immediate contribution to the reduction of fossil-fuel use and carbon-dioxide emissions, at least in the United States.

The two key obstacles confronting the US nuclear industry concern safety and cost. Ever since the Three Mile Island accident in March 1979—in which, for the first time, a reactor core partially melted and small amounts of radioactivity were released into the atmosphere—the perception by many people that nuclear plants are unsafe has inhibited plans to build additional plants. According to a Louis Harris poll conducted in late 1988, 61 percent of Americans are opposed to the construction of more nuclear plants, while 30 percent are in favor—almost exactly the reverse of opinions found in answer to the same question 10 years earlier (Harris Poll 1989).

At the same time, electric utilities and private investors have become disenchanted with the rapidly rising costs of nuclear power, once predicted to be "too cheap to meter." (See Figure 7.) Nuclear plants completed in recent years have cost an average of $2,700 per installed kilowatt of capacity, not including $800-900 per installed kilowatt for real interest charges incurred during construction. Such high capital costs, combined with substantial operations, maintenance, and fuel costs, make electricity generated by new nuclear plants almost twice as expensive, on average, as that generated by new coal-fired plants (Komanoff 1988).

The two issues, cost and safety, are not unrelated. Changing safety regulations requiring expensive design modifications and backfits have added to the cost of new reactors. However, the nuclear industry was experiencing problems of low performance and escalating costs well before the Three Mile Island accident (Komanoff 1981). Although a complex regulatory environment is often blamed for the industry's problems, poor management is the more plausible culprit (Hansen *et al.* 1989).

Development of New Reactor Designs

With the existing light-water reactor (LWR) designs at an impasse, attention has been focused on the possible development of safer and more economical reactor designs to replace them. Some companies are concentrating on simplifying and standardizing LWR designs to improve their performance and safety (Stricharchuk 1989, Electrical World 1989). Three quite different types of reactors are also under development in the United States and abroad: the Process Inherent Ultimate Safety (PIUS) reactor, of Swedish invention; the modular gas-cooled reactor (MGR), developed by American and West German firms; and the Integral Fast Reactor (IFR), based on a design by the Argonne National Laboratory. All of these designs depend to some degree on safety features requiring comparatively little human or mechanical intervention to prevent, or at least delay, core meltdown and the release of radioactivity. They are also relatively small reactors, with capacities of 100-300 megawatts (MW), in contrast to the 1,000 MW typical of today's reactors. Smaller reactors, in theory, would require less capital investment and risk and could be tailored more closely to electricity demand (Pollard 1987, Lidsky 1987, Broad 1988).

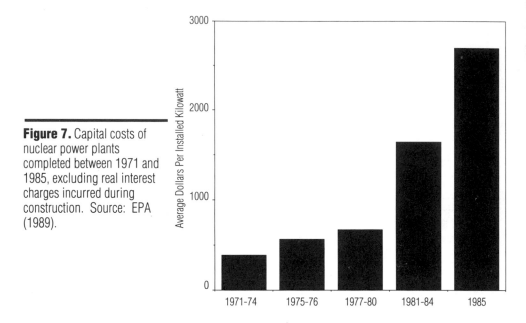

Figure 7. Capital costs of nuclear power plants completed between 1971 and 1985, excluding real interest charges incurred during construction. Source: EPA (1989).

While the advanced reactor designs deserve study, however, none has yet been demonstrated on a commercial scale, and it will be several years before their true safety and commercial viability are determined. Even then, public acceptance is not guaranteed.

Storage of Radioactive Wastes

Whether or not advanced reactor designs prove successful, other important problems call into question the feasibility and wisdom of greatly expanding reliance on nuclear power. All nuclear power plants produce hazardous, radioactive wastes, which must be safely stored for up to thousands of years. At present, most plants store their wastes on site, but this is a temporary solution. The United States has initiated investigations for a permanent waste repository to be constructed underground in Yucca Mountain, Nevada. The repository, which will hold wastes generated by existing commercial plants, is scheduled to be opened in 2010 at the earliest.

The Yucca Mountain project has already been delayed several years, however, and is facing growing opposition from Nevada citizens. The much smaller Waste Isolation Pilot Project (WIPP) in New Mexico, intended to test the concept of deep-underground storage using wastes from nuclear-weapons production facilities, has also been delayed (Reid 1989). Questions concerning the danger of water intrusion into salt caverns, which could result in the eventual contamination of local groundwater, have not been resolved. Given the engineering difficulties and mounting political problems associated with siting waste facilities, it is uncertain whether new sites could be found to store wastes generated by plants constructed in the future.

Nuclear-Weapons Proliferation

An additional risk of expanding reliance on nuclear power is the possibility that nuclear materials will be diverted by countries or terrorist organizations for the purpose of making nuclear weapons. The risk of such diversions from the United States is presently quite low, because the "once-through" uranium fuel cycle of commercial reactors does not generate materials directly usable in weapons. However, because of limited US and world uranium reserves, a world energy economy relying heavily on nuclear power would require the reprocessing of spent fuel and the construction of breeder reactors to produce plutonium. In contrast to low-enriched uranium, plutonium can be readily used to make nuclear weapons. In a nuclear-based economy, large quantities of plutonium fuel would have to be transported on roads and railways and across oceans, posing a potentially serious risk of material diversion (not to mention accidents). It was in part because of this danger that the United States decided to forgo fuel reprocessing in the 1970s (Lipschutz 1980).

Summary

For at least the next two decades, nuclear power plants cannot be counted on to supply much more energy in the United States than they do today, despite efforts by the Bush administration to lead a revival of the nuclear industry (Victor 1989a). Indeed, their contribution may begin to decline after the turn of the century as aging plants are retired, if new plants are not ordered in the near future. Nuclear power may fare somewhat better in other countries, but probably not much, except in France and a few other countries where public opposition remains limited. The 1986 Chernobyl disaster in the Soviet Union has sparked a reevaluation of the role of nuclear power in the energy plans of many countries, and this is reflected in a marked worldwide slowdown of plant construction (EIA 1988a). While a greater role for nuclear power cannot be ruled out in the long term, this will require the resolution of serious problems concerning the safe disposal of radioactive wastes and the prevention of nuclear-weapons proliferation.

Renewable Energy Sources

Renewable energy sources[2] are often regarded as new or exotic, but in fact they are neither. Until quite recently, in historical terms, the world drew most of its energy from the sun, either directly from sunlight or indirectly through the natural processes that generate winds, rivers, and plants. In the 19th century, the most common source of energy in the United States was firewood. In areas where streams were plentiful, water power was used to thresh grain and mill lumber, and on farms windmills for pumping water were a common sight. Houses were often oriented to capture sunlight in winter and provide shade in summer, and storage tanks were painted black so that water would be heated by sunlight.

Most of these practices were gradually abandoned as fossil fuels—first coal, then oil and natural gas—came into wide use. The two major exceptions are hydroelectric power, which now supplies 10-12 percent of the electricity generated in the United States, and the combustion of wood and wood wastes by homes and industries, which accounts for 4-5 percent of primary-energy consumption. (Geothermal energy, which currently provides about 0.25 percent of the US energy supply, is not discussed here.)

Interest in renewable energy sources revived in the 1970s out of concern over US dependence on oil imports and pollution caused by fossil-fuel use. Federal funding for renewable-energy research and development (R&D) rose quickly, from $75.1 million in fiscal 1975 to $718.5 million in fiscal 1980 (Sissine 1989). A variety of new technologies were developed and old ones improved, and some attained a measure of commercial success; wind turbines, for example, were installed in large numbers in California in the early 1980s. As the sense of energy crisis faded, however, so did interest in developing alternatives to fossil fuels. R&D funding steadily declined, reaching a low of $114.7 million in fiscal 1989—a cut of almost 90 percent below the 1980 level, if inflation is taken into account.

Yet the advantages of renewable energy sources—particularly wind, solar, and biomass (plant) energy—are, if anything, more compelling today than ever before. The technologies that have been developed, ranging from wind turbines and photovoltaic cells to liquid fuels derived from biomass, are of startling versatility. Most produce little or no pollution and hazardous wastes. Drawing entirely on domestic resources, they are immune to foreign disruptions like the 1973 Arab oil embargo, and since their "fuel"—sunlight—is free, they provide a hedge against inflation caused by the depletion of fossil-fuel reserves. Their development would almost certainly result in a net increase in employment, as renewable-energy industries generally require more labor, per unit of energy produced, than coal, oil, and natural-gas industries (Hall *et al.* 1986).

Most important, resources of renewable energy are vast and inexhaustible. Sunlight falling on the US landmass carries about 500 times as much energy as the United States consumes in a year. Wind, biomass, and hydroelectric resources, though more modest, are also substantial. In practice, of course, only a fraction of these resources could be exploited because of constraints on available land, the efficiencies of energy conversion, and other factors. Nevertheless, our estimates indicate that the physically recoverable resources would more than suffice to meet current and foreseeable energy demand (see Table 2). Solar energy has by far the greatest potential: solar collectors covering less than 1 percent of US territory—one-tenth the area devoted to agriculture—could make more energy available than the United States consumes in a year. Hydroelectric power has the least room for further expansion, since a large portion of the river resource has already been developed and much of the rest is barred from development by federal environmental legislation. (For further details and references, see the chapters on individual technologies.)

Technologies and Economics

Despite the impressive potential of renewable energy sources, they have been virtually ignored by most mainstream energy analysts, many of whom regard them as expensive and impractical. Yet the costs of renewable-energy technologies have declined dramatically in the 1980s, and their reliability has been proven in government/industry demonstration projects and actual commercial operation. For some emerging technologies, all that is needed to become fully competitive is a market demand large enough to justify economies of scale. For others, further research and development are required, but their long-term prospects are bright.

Wind turbines are a good example of the growing competitiveness of renewable-energy technologies. The cost of electricity generated at windy sites has declined from over 25¢/kWh in 1981 to 7-9¢/kWh today, and it could fall as low as 4-6¢/kWh in five years.[3] At the current price, wind power is competitive, or nearly so, with electricity generated by new fossil-fired power plants, and by the early 1990s it should be one of the least expensive sources of electricity, fossil or renewable (CEC 1988a). Reliability problems affecting early wind-turbine designs have been largely resolved, and mature and well-maintained systems are available 95-98

Table 2. Estimates of the theoretical and recoverable resources of renewable energy in the United States and their current utilization, in exajoules per year. The United States consumes approximately 84 exajoules annually. For references, see individual technology chapters.

Energy Source	Theoretical Resource	Recoverable Resource	Current Utilization
Solar	46,000	>100	0.06
Wind	3,000	10-40	0.02
Biomass	—	13-26	3.11
Hydroelectric	5.8	3-4	3.24

percent of the time. Other renewable sources of electricity, such as solar-thermal electric-power plants and photovoltaic cells, also promise to become competitive within a decade, particularly if market demand grows large enough to permit economies of scale (see Figure 8).

For applications requiring direct heat—over half of the end-use energy consumed in the United States—solar-thermal systems are becoming more attractive. Systems now on the market designed for commercial and industrial use generate hot water or steam in sunny regions at about 1.5-3 times the current cost of production with natural gas. If the market for solar-thermal systems were larger, the cost of energy could fall sharply. Passive-solar building designs, which use a building's structure to capture and store solar energy, can greatly reduce energy consumption for space heating, cooling, and lighting for little extra cost when incorporated into new construction.

Developing renewable substitutes for gasoline and other transportation fuels is perhaps the most difficult challenge, but even here there is promise of a near-term solution. Ethanol can now be produced from wood and other plants at about twice the cost of conventional gasoline (without taxes and distribution). With continued improvements in conversion processes and the cultivation of biomass feedstocks, and with expected increases in the cost of gasoline, the two fuels could become roughly competitive around the turn of the century (see Figure 9). According to our estimates, biomass fuels, including ethanol, methanol, and plant oils, could ultimately power 30-90 percent of cars and trucks in the United States. Forestry and agricultural wastes, as well as plants and trees grown specifically for energy, would supply the raw materials. Further in the future, cars powered by hydrogen or even electricity (provided by low-cost renewable sources, such as photovoltaic cells) are a realistic possibility.

Energy Storage

What happens when the sun goes down or the wind stops? Conventional wisdom holds that energy storage will be needed to keep the power flowing reliably, substantially raising the costs of solar and wind power. But while the variability of wind and solar power is an important issue, it should not hinder the commercialization of these technologies in the near term. In some applications, considerable storage or backup capacity already exists. For example, electric utilities maintain a reserve capacity (typically 20 percent of peak demand) to allow for plant shutdowns. This reserve should suffice until solar and wind energy constitute at least a few percent, and possibly more than 20 percent, of the total electricity supply—a level of market penetration that will not be achieved for at least a decade.

Furthermore, hybrid energy systems drawing on both renewable and fossil sources could provide reliable power while greatly reducing fossil-fuel consumption. Power plants have been built in Southern California, for example, which run on 75 percent solar energy and 25 percent natural gas and supply reliable peak power year-round. Natural gas could also supplement solar energy in residential, commercial, and industrial heating applications for little extra cost. By the time solar and wind energy come into wide use, more cost-effective storage technologies are likely to be available.

Costs of Electricity

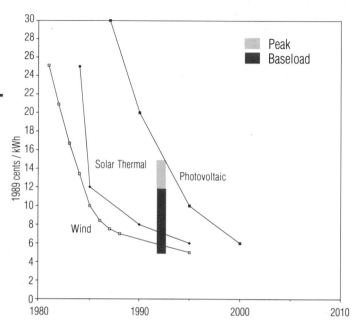

Figure 8. Trends in the cost of electricity generated from solar-thermal, photovoltaic, and wind energy sources; for references, see solar- and wind-energy chapters. Bar represents projected range of "acceptable" baseload and peak electricity costs for independent power producers in California in 1992, adapted from CEC (1988a).

Costs of Ethanol and Gasoline

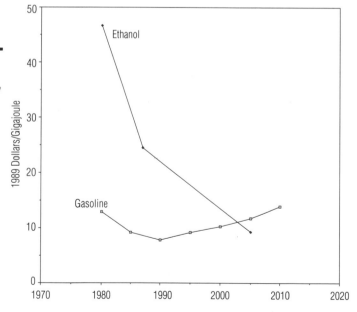

Figure 9. Comparison of historical and projected costs of ethanol derived from woody biomass with conventional gasoline. Sources: DOE (1988a, 1988d).

Future Prospects

For all of the progress made by renewable-energy technologies in the past decade, many are still struggling for commercial acceptance. A glut of low-cost fossil fuels, sharp cutbacks in federal support, and above all an energy market obsessed with short-term profits and insensitive to the environmental and social costs—such as pollution and global warming—of conventional energy technologies, have conspired to slow industry growth in the latter half of the 1980s. According to official forecasts, renewable energy sources are likely to be providing only about 9.5 percent of projected US energy supply in 2000 and 12 percent in 2010, compared to 7.6 percent today (DOE 1988a). (See Chapter 7.)

But with appropriate market incentives and strong government leadership, this picture could change dramatically. We estimate that the share of US energy supply provided by renewable energy sources could be increased to 15 percent—double the present fraction—by 2000, resulting in a 5-10 percent reduction in fossil-fuel use and carbon-dioxide emissions. We further believe that as much as 50 percent of US energy supply could be provided by renewable energy sources by 2020, assuming no overall growth in energy consumption.

Although achieving such a high rate of increase in renewable-energy consumption is admittedly an ambitious goal, it is not without historical precedent. For example, oil consumption rose from 6 percent of total energy use in 1910 to 23 percent in 1930—a factor of four increase in 20 years. Similarly, natural-gas consumption increased from 11 percent in 1940 to 29 percent in 1960 (Kendall and Nadis 1980). Much of the growth in renewable-energy consumption before 2000 would probably come from greater exploitation of biomass resources; by 2020,

Table 3. One possible scenario of renewable-energy supply in 2000 and 2020 (in exajoules per year), assuming implementation of policies to encourage development of renewable energy sources.

		Year	
	1989	**2000**	**2020**
Solar	0.06	1.5	20.0
Biomass	3.11	7.0	15.0
Wind	0.02	0.5	5.0
Hydroelectric	3.24	3.5	3.5
Total	6.43	12.5	43.5
Percent of 1989 Consumption	7.6%	14.9%	51.7%

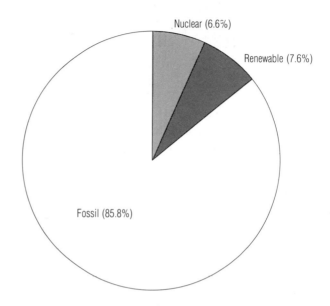

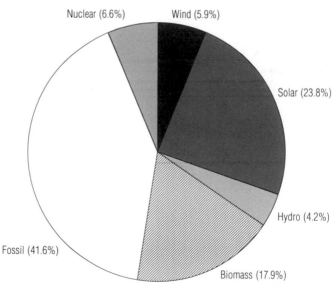

Figure 10. US primary-energy supply in 1989 (above) and projected to 2020 (below). Excludes geothermal energy (current contribution 0.25 percent). Assumes no overall growth in energy consumption; total 84 exajoules. Source: Table 3.

however, solar energy would probably be the dominant source (see Figure 10). Most initial solar and wind development would occur in areas with abundant sunshine and wind, but as the costs of these technologies decline through technological improvements and economies of scale, development should spread into other areas as well.

Mitigating Global Warming

Could energy efficiency and renewable energy sources play a major part in helping to slow and eventually stop global warming? The answer appears to be yes. If nations took full advantage of opportunities to improve energy efficiency, global fossil-fuel use and carbon-dioxide emissions would grow slowly, if at all. And if, in addition, renewable energy sources were developed to their full potential, fossil-fuel use and carbon-dioxide emissions could be cut well below today's levels, eventually approaching the 50-80 percent reduction necessary to stabilize the global climate.

This was the conclusion of an Environmental Protection Agency study that examined a wide array of policy options for mitigating global warming (EPA 1989). Under the EPA's rapidly-changing-world (RCW) scenario, policies designed to promote energy efficiency and renewable energy sources were judged able to reduce the earth's warming commitment (temperature increase to which the earth will be committed) an estimated 27 percent by 2050 and 42 percent by 2100. Energy efficiency and renewable energy sources were about equally important in this analysis and together accounted for nearly two-thirds of the total warming reduction—65 percent—deemed feasible in the next century. In contrast, nuclear power's potential contribution was judged to be less than one-third that of either energy efficiency or renewable energy sources alone. (Among other policies considered were changes in agricultural practices and reductions in the use of chlorofluorocarbons.)

One of the most important conclusions of the EPA study was that renewable energy sources could make up 30-45 percent of the projected global primary-energy supply by 2050. In fact, we believe that they could be developed even more quickly in the United States, because of the great technical and financial resources concentrated here. Starting almost immediately, the United States could begin following a transitional path to an energy economy relying almost entirely on renewable energy sources. However, this will only happen with government leadership dedicated to removing the many barriers, described in Chapter 7, hindering the development of these sources.

Notes

1. Switching from coal and oil to natural gas would also reduce carbon-dioxide emissions; however, domestic and world gas reserves are limited, so that fuel-switching would be of benefit only in the near term.

2. Renewable energy is defined as energy that is replenished from a virtually inexhaustible source at a rate equal to the rate at which it is consumed. Energy in sunlight, the winds, rivers, and plants is consequently renewable, as is energy in tides created by the gravitational pull of the moon and sun. Geothermal energy, on the other hand, is not renewable (with the possible exception of the thermal energy in magma) because of the slow rate of heat transfer from the earth's interior to the surface. Geothermal resources near the surface are nevertheless theoretically very large (Kendall and Nadis 1980).

3. Unless otherwise noted, electricity costs are in 1989 dollars levelized over 30 years. Care must be taken when comparing costs, as assumptions (such as fixed-charge rate) may differ.

Chapter three

Solar Energy

In Aeschylus' play *Prometheus Bound*, the god Prometheus recounts how he found the people of earth unenlightened and lacking "the knowledge of houses turned to face toward the sun." As this passage suggests, the ancient Greeks had a deep appreciation of the value of solar energy—an appreciation that our civilization, for all of its technological achievements, lacks.

Yet since Aeschylus' time, a variety of new technologies for converting sunlight into useful energy have been developed and old ones improved. In the past decade, about 250,000 houses have been built with passive-solar designs, which use a building's structure to capture and store heat, and over two million residential solar water-heating and space-heating systems have been sold. Concentrating solar collectors—curved mirrors or lenses that focus light onto a receiver—have been marketed for generating industrial heat, steam, and electricity. And photovoltaic cells, which convert sunlight directly into electricity, have become a $100-million industry for applications requiring power in remote areas and for consumer products.

At the same time, solar technologies have become far more cost effective than they were only a decade ago. Solar-thermal electric systems now generate power for as little as 12¢/kWh, and the cost is expected to fall to 8¢/kWh when larger systems under construction are completed. Photovoltaic cells are 10 times less expensive than they were a decade ago, and they, too, could become competitive in electric-utility markets in the near future. Passive-solar building designs are cost effective today for new construction, and design and material improvements will make them more so in coming years.

Despite such progress, solar energy supplies only a tiny fraction of US energy demand—no more than 0.05-0.075 percent, or 0.04-0.06 exajoules (EJ) per year (Sklar 1989, DOE 1988a). In recent years, the market for solar energy has shrunk because of low fossil-fuel prices and the elimination of federal renewable-energy tax credits, while the pace of technological improvement has slowed because of sharply reduced federal funding for research and development. Without additional market incentives and strong government leadership, solar energy is unlikely to become a substantial source of energy in the United States for at least another decade and possibly longer.

The Solar Resource

Solar energy boasts the largest resource potential of any renewable or nonrenewable energy source. According to one estimate, over 46,000 EJ (44,000 quads) of sunlight fall on the US landmass each year, an amount equivalent to about 500 times US primary-energy consumption of 84 EJ (Kendall and Nadis 1980). The solar resource is unevenly distributed geographically, but differences between regions are not as great as one might expect. In the Northeast and Northwest, the average daily insolation (incident energy per unit area) is about 14 megajoules per square meter (MJ/m^2), while in the sunny South and Southwest it is 50-100 percent higher.

The amount of sunlight that could be collected and converted into energy is constrained by the amount of available land and the efficiency of energy conversion. Even so, it appears that more than enough solar energy could be generated to satisfy our needs. If just 1 percent of US land area were adapted to collecting solar energy at 25 percent conversion efficiency, over 100 EJ would be made available each year. The amount of land area required to generate electricity from sunlight is, in fact, comparable to the amount of land required to generate it from coal when the land disturbances of coal mining are taken into account (Meridian 1989). Moreover, solar collectors and photovoltaic modules can be mounted on otherwise unused space on the roofs of houses and buildings.

Technologies and Economics

Solar technologies are usually divided into three categories: solar buildings, solar-thermal collectors, and photovoltaic cells. The solar-building category includes active and passive water heating, space heating, and space cooling. Solar-thermal collectors concentrate sunlight to temperatures (100-1,500°C) that can be used for industrial processes and to generate electricity. Photovoltaic systems produce electricity directly through the interaction of sunlight with semiconductor materials.

Solar Buildings

Solar buildings have received less attention and funding than other solar technologies, but their potential energy contribution is large. The residential and commercial sectors of the economy account for approximately one-third of US energy use (including energy used to generate electricity). The residential sector alone accounts for almost 20 percent of energy use, and of that two-thirds goes for water heating, space heating, and space cooling (EIA 1989). If solar-building technologies were widely adopted in new construction, they could plausibly displace up to 20-25 percent of US fossil-fuel use over the next 50 years (Kendall and Nadis 1980). Solar buildings currently displace a very small amount of fossil energy—0.055 exajoules (0.052 quads), according to Sklar (1989).

Solar-building technologies are characterized as either passive or active. Passive technologies use a building's structure to capture sunlight and store heat, reducing the requirements for conventional heating and lighting. Methods include window glazings that admit incoming sunlight but trap outgoing infrared radiation, convection systems that circulate heated air or water through a building by natural convective forces, large masses built into a structure that store heat from the sun and release it slowly at night, and simple architectural considerations, like orienting buildings to take maximum advantage of sunlight. Buildings can also be cooled by ventilating interior spaces, evaporating water, and using the earth underground to absorb heat. In addition, windows and skylights can be designed to supplement light for interior spaces.

When designed in from the start, passive technologies need not add significantly to the construction cost of a home or building, although retrofitting existing structures can be expensive (IEA 1987). Current technologies, moreover, appear to be quite effective. A survey of 38 passive solar homes conducted by the Department of Energy indicated average savings of almost 40 percent in the amount of heating required (DOE 1988e). According to estimates by the National Association of Home Builders, the extensive use of passive-solar designs could reduce a typical building's heating, cooling, and lighting needs by 50-60 percent while adding 10-12 percent to construction costs (cited by Rader 1989). This performance could improve further—to the point where 80 percent of a building's heating needs and 60 percent of its cooling needs could be met through passive-solar designs—with the implementation of new technologies under development, including phase-change materials that store heat more efficiently than water, rocks, and other ordinary materials; improved thermal glazings for windows and walls; light pipes to bring sunlight into interior spaces; and electrochromic window films that admit or block sunlight according to a minute electrical current (DOE 1989a, Rosenfeld and Hafemeister 1988).

The cost of passive solar technologies, though difficult to determine, appears to be in the range of $0-5 per gigajoule (GJ) of energy saved, depending on the particular technology, local climate, and other factors (see, e.g., Rader 1989). For example, double-paned, low-emissivity windows are estimated to cost approximately $4/GJ of energy saved, and the cost is expected to fall to $2/GJ as the market matures (Rosenfeld and Hafemeister 1988). By comparison, residential consumers paid an average of $4.26/GJ for natural gas and $17.60/GJ for electricity in 1988 (EIA 1989a).

Despite the economic and aesthetic attractiveness of passive-solar technologies, they have not been adopted by the construction industry as rapidly as one might expect. In the United States, about 250,000 passive-solar homes have been built (DOE 1989a), but this is a small number compared to the total number of households (80 million); only a few nonresidential solar buildings are in existence. The slow progress can be attributed in part to the time it takes for new designs and building techniques to infiltrate the industry. The Department of Energy, the National Association of Home Builders, and the Passive Solar Industries Council have recently cooperated to make passive-solar design guidelines more widely available to home builders (DOE 1989a, NAHB 1988). Even with such aids, however, firms often have little incentive to incorporate passive-solar designs into buildings they design and build, since they do not pay the costs of heating and cooling the buildings in later years. Owners and tenants, for their part, often demand an unrealistically short payback period (as little as two to three years) for investments in passive-solar technologies and other energy-saving measures.

Active-solar systems are discrete units that collect, store, and distribute solar energy for water heating, space heating, and space cooling. Many types of solar collectors have been developed, the simplest of which is the flat-plate collector. In a typical design, a black metal plate absorbs sunlight and transfers the heat to pipes carrying water or a water-alcohol mixture (to prevent freezing). The absorber plate is topped by glass, which is sometimes glazed to minimize radiative heat loss, and the rest of the system is surrounded by insulation. Over the years this simple design has been improved to the point where, on an average basis, up to 50 percent of incident solar energy is transferred to the water, and peak efficiencies can be as high as 80 percent (IEA 1987). More advanced designs include the evacuated-tube collector, in which the insulation of an ordinary flat-plate collector is replaced with a vacuum to minimize heat losses, and the concentrating collector, usually a parabolic-trough reflector, which focuses sunlight onto a fluid-carrying tube, producing higher temperatures than the flat-plate design.

In water- and space-heating applications, the heat from the solar collectors is usually transferred to an insulated storage medium (such as a water tank) and distributed as needed. In addition, the heat can drive a cooling system. A variety of cooling technologies are being studied, ranging from conventional refrigeration cycles to desiccant-evaporators, which absorb moisture from the air and then evaporate it. Solar cooling systems are not yet very effective, though their performance should improve in time. Evacuated-tube and concentrating collectors should provide the higher temperatures needed for more efficient operation, and more effective desiccant materials are under development.

The US market for solar collectors has had a roller-coaster ride, growing quickly in the early 1980s and then collapsing just as quickly with the decline in fossil-fuel prices and the end of residential renewable-energy tax credits in 1985 (see Figure 11). In 1984, over 200 manufacturers sold about 1.5 million square meters of solar collector (both flat-plate and concentrating). By 1987, about 50 manufacturers were left and sales were down to less than 0.4 million square meters (EIA 1989b). Even at the market's peak, solar collectors never played a significant role in the nation's energy supply. Most of the over two million collectors installed in American homes were for heating swimming pools.

Sales of Solar Collectors

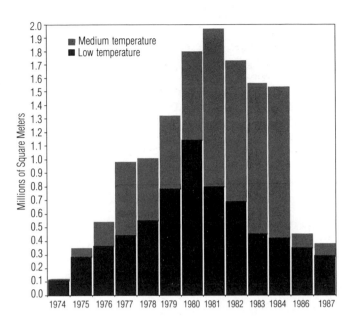

Figure 11. Sales of low- and medium-temperature solar collectors, in millions of square meters, 1974-1987. No data were gathered in 1985. Source: Energy Information Administration (1989b).

This experience has been echoed in Europe and other parts of the world, with some notable exceptions. In Cyprus, over 90 percent of homes are now equipped with solar water heaters, as are over 65 percent of homes in Israel. In fact, Israeli law requires all new residential buildings up to 9 stories high to use solar water heating. In parts of Australia, the market penetration is over 30 percent (Shea 1988, IEA 1987).

The high initial cost of active-solar systems—ranging from $2,000 for small water-heating systems to over $8,000 for residential space-heating systems (DOE 1989a)—is the main reason for low sales in the United States. Cooling systems are even more expensive, costing $10,000-$20,000 for typical homes. The most a solar water-heating system could be expected to save a homeowner is about $500 per year when the competing fuel is electricity[1] and much less when it is natural gas. Thus the simple payback period is at least four years, longer than that demanded by most consumers. When calculated over the lifetime of the collectors, however, solar water heating appears to be cost effective, in sunny climates, in comparison to electric water heating.

In other countries, solar water-heating systems are considerably less expensive, implying that the price could be lower in the United States as well. In some countries the price per square meter is one-half that in the United States (IEA 1987). In Israel, the average price of installed systems is estimated to be $500, and

31

the payback period is less than four years (Shea 1988). Research in the United States is focusing on making solar collectors out of lightweight plastics and developing more efficient and higher-temperature designs, such as the evacuated-tube collector. Probably the main requirement to achieve lower costs, however, is a market demand large enough to justify economies of scale. According to solar-collector manufacturers, as much as a third of the price of solar water-heating systems goes to pay for marketing, an indication that the industry is still struggling for consumer recognition (IEA 1987).

Space heating has accounted for only a small fraction—10 percent or less—of collectors sold. The main reason is that, on an annual basis, space heating is less efficient than water heating because it is not needed during warmer months when insolation is highest. In one test program, commercial water heaters were found to be up to 41 percent efficient on an annual basis while space heaters were at most 22 percent efficient (IEA 1987). The cost-effectiveness of space heating could be greatly improved by storing summer heat for winter use (discussed below).

Solar-Thermal Systems

The heat from solar collectors, when concentrated to moderate and high temperatures, can be used in a great number of industrial applications and to generate electricity. The potential market for solar-thermal energy is consequently very large: industrial use of fossil fuels accounts for about one-quarter of US energy consumption, and electric utilities account for an additional one-third (EIA 1989b). Some inroads have already been made in the electricity market. In the past five years, almost 200 megawatts (MW) of solar-thermal electric capacity have been installed with private capital in California, and an additional 400 MW are planned by 1992. Industrial process heat applications remain largely undeveloped, and today solar-thermal systems generate the primary-energy equivalent of just 0.005 EJ (Sklar 1989).

There are four main types of solar-thermal collectors: the parabolic trough, parabolic dish, central receiver, and solar pond. A parabolic trough is a reflector that focuses sunlight onto a line receiver, usually a vacuum-enclosed metal or glass pipe, in which a fluid is heated to a temperature of up to about 400°C. Troughs are usually mounted on a north-south axis and track the sun. A parabolic dish consists of a bowl-shaped reflector that focuses sunlight onto a small area, achieving a temperature as high as 1,500°C, though lower temperatures are more common. The dish is usually mounted on a two-axis sun tracker. In a central-receiver system, a large array of sun-tracking mirrors, called heliostats, focuses sunlight onto a receiver mounted on a tower. A solar pond is a body of saline water that traps heat between layers of varying salt density. (Ponds are likely to be feasible in only a few locations and will not be discussed further here.)

Whatever the system, the collected heat may be used directly or to generate steam and electricity. Both the trough and dish designs have the advantage that they are modular: each unit is relatively small (25-100 kilowatts peak thermal capacity is typical), so that the size of the entire system can be scaled to suit almost any application. Central-receiver systems are generally much larger—at least 30 MW—to take advantage of economies of scale.

Parabolic-trough systems are well-suited to producing heat for industrial processes, approximately half of which require temperatures less than 300°C. Several solar industrial-process-heat plants were built in the early 1980s for such diverse customers as a tractor company, a commercial laundry, a brewery, and a dyeing factory (IEA 1987). In recent years, however, attention has shifted to generating electricity.

Luz International Ltd. is the clear leader in the development of parabolic troughs for electricity generation. Seven Luz Solar Electric Generating Systems (SEGS) are operating in California, the first built in 1985 and the last completed at the end of 1988. Each plant (except the first) generates up to 30 MW of electricity for sale to Southern California Edison. The plants consist of a large array of parabolic troughs through which passes a fluid, which is then piped to a central plant where steam and electricity are generated. According to Luz, the levelized cost of electricity produced by the most recent SEGS is just under 12¢/kWh (Luz 1989). Five new and larger (80 MW) Luz plants are planned, the first to be finished in 1989 and the last in 1992, and the cost of electricity from those plants is projected to be less than 8¢/kWh (see Figure 8).

Several factors are responsible for Luz's success. First, the 1978 Public Utilities Regulatory Policies Act (PURPA) and federal and state tax incentives secured an initial utility market for Luz plants even at their high initial cost (Cook 1989). Second, Luz has succeeded in greatly reducing capital costs and improving performance since the first plant. The company reports that since 1984 the installed cost of collectors has fallen 75 percent to $130-$180/m² (Harats and Kearney 1989) and now comprises less than one-third of the total capital cost of $2,500 per installed peak kilowatt. In addition, the newer Luz plants produce electricity 20 percent more efficiently than earlier plants, thanks to higher temperatures achieved in the collector fluid (Jensen, Price, and Kearney 1989).

Third, and perhaps most important, the Luz plants are well suited to meeting peak summer daytime needs in the Los Angeles area, for which Southern California Edison pays a large premium. Even though only 18 percent of annual SEGS output is generated in peak periods, over 45 percent of annual revenues are earned then (Jensen, Price, and Kearney 1989). Furthermore, natural gas is used to supplement solar energy, allowing the SEGS to generate power reliably when its value is highest. (The amount of energy annually derived from natural gas is restricted by PURPA regulations to 25 percent of the total output from a solar installation.)

Although Luz Industries has found a temporary niche in which to grow, its long-term future depends on reducing costs still further. The cost of electricity could be reduced to about 6¢/kWh simply through the construction of larger plants (100-200 MW capacity) than those permitted under PURPA regulations. In addition, company researchers are seeking a 20-30 percent cost reduction through changes in plant design (such as direct steam generation in the collectors). Achieving a cost of 6¢/kWh would make this technology highly competitive with conventional electrical-generation technologies in the peak-power market.

Parabolic-dish systems produce higher temperatures than parabolic troughs, making them suitable for such industrial applications as the production of metals, glass, cement, paper and brick, as well as more efficient electricity generation than is possible with parabolic-trough systems. Most existing parabolic-dish systems are similar in concept to the Luz SEGS in that a heat-transfer fluid is passed through the receivers of the dishes and piped to a central location before being used. Another concept that may be especially suitable for small-scale, remote applications is to equip each parabolic dish with its own electricity generator. One model that has been developed, the Vanguard Dish-Stirling module, produces up to 25 kilowatts of electricity at a remarkably high net solar-to-electric conversion efficiency of about 29 percent.

The cost of parabolic-dish systems has been dramatically reduced over the past ten years or so. Systems once priced at over $1,000 per square meter installed are now estimated to cost about $150 per square meter installed (DOE 1989b). The latter figure implies a levelized cost of thermal energy, in sunny regions, of about $7-$9/GJ.[2] Dish-electric systems are estimated to cost approximately $3,000 per installed kilowatt of capacity, or slightly more than the latest Luz plants, and could generate electricity at about 10¢/kWh (DOE 1989b). The costs are expected to decline further in the 1990s. The Department of Energy has set a 1995 goal of $1,000 per installed kilowatt for large systems, at which point electricity could be sold for 5¢/kWh (DOE 1989b). This goal may be achieved with higher-temperature receivers (operating at up to 1,400°C) and stretched-membrane reflectors, which are lighter and potentially less expensive than conventional glass-metal reflectors. Again, however, as with parabolic-trough systems, achieving economies of scale is probably the most important prerequisite to commercial success. If produced commercially today, the Vanguard Dish-Stirling module might cost $4,000 per installed kilowatt, but some believe that a 50 percent price reduction could be achieved through mass production (IEA 1987). Major improvements in the durability of the Stirling engines that are at the heart of this concept will also be required for successful commercial application.

Despite the apparent promise of parabolic-dish systems, no significant market for them has appeared. Most prototypes have been sold to governments and utilities for individual testing, with the exception of a 4.9 MW (electric) plant built in 1984 by LaJet Incorporated in Warner Springs, California, and a 3 MW (thermal) plant built in 1982 and operated by the Georgia Power Company in Shenandoah, Georgia. (The Georgia plant generates electricity plus steam for a knitwear factory.) The Warner Springs facility is to be taken over and refurbished with stretched-membrane reflectors manufactured by Science Applications International Corporation (Porter 1989).

Central-receiver systems were the initial focus of solar-thermal research, though enthusiasm for the concept has waned somewhat recently in the United States. Several experimental plants were built in the early 1980s—including the largest in the world, the 10 MW Solar One plant near Barstow, California—in the United States, France, Italy, Spain, and Japan, and in 1986 one was built in the Soviet Union (Shea 1988). These plants have given researchers valuable field experience with a variety of central-receiver designs. With the decline in government funding in recent years, however, US researchers have turned their attention to more basic research on system components, in particular heliostats, receivers, and thermal transport and storage systems (some of which have application in parabolic-dish design as well). Interest remains stronger overseas. A 30 MW central-receiver project under the direction of a European consortium is expected to be completed in Jordan by 1994. The Soviet Union also plans a 300 MW system with a natural-gas backup (Porter 1989).

Central-receiver plants are currently estimated to cost about $3,000 per installed kilowatt, making them roughly comparable in capital cost to parabolic-dish systems. Their annual energy-conversion efficiency is lower, however, resulting in a higher cost of electricity produced (DOE 1989b). The Department of Energy expects to reduce the capital costs of central-receiver systems by over 50 percent within the next decade with stretched-membrane reflectors. The efficiency of the receivers could also improve with the use of liquid sodium or molten salts to capture and transport heat.

With these and other advances as well as the construction of larger plants, central receivers could probably be competitive for generating peak power by the mid-to-late 1990s. One study by Bechtel National and Pacific Gas and Electric estimated that central receivers of 100-200 MW size could be producing electricity at a cost of 8-11 ¢/kWh by 1997 (PG&E 1988). However, the relatively large size of these systems—which are not by nature modular—may limit the initial market for them in the United States (Marshall 1989).

Photovoltaic Cells

Photovoltaic cells are the most elegantly simple solar-electric technologies, and perhaps for that reason they have received the most public attention. Today there is a small but growing photovoltaic industry supplying cells for many applications, ranging from calculators to water pumping. Major cost reductions and performance improvements will be needed for photovoltaic electricity to begin entering the utility-grid market, but with more intensive research and development there is a good chance that this goal can be met within the next few years.

In photovoltaic cells, electricity is generated by photons of light knocking electrons loose from atoms (see Figure 12). The negatively charged electrons and positively charged "holes" are swept onto opposing metal contacts by a voltage created between two semiconductor materials. When the circuit is closed, an electric current is established. Since 1954, when the first modern photovoltaic cell was demonstrated, a vast array of materials and designs has emerged from government and private research, and new developments are occurring all the time.

Silicon (Si) has long been the dominant material for photovoltaic cells because of its well-known properties and wide use as a semiconductor. Single-crystal silicon cells were the first to be developed and still hold just under half of the market (IEA 1987). Among silicon cells, single-crystal offers the highest efficiency of conversion from sunlight to electricity. (Efficiency is a key factor in cost, since more efficient cells can produce more electricity using a smaller cell or collector area.) In 1988, researchers set a new record when they measured an efficiency of 22.8 percent in a small single-crystal cell under normal (unconcen-

Figure 12. Schematic drawing of a typical photovoltaic cell. An electric field is created at the junction of two dissimilar semiconductor materials, sweeping electrons and "holes" onto opposing electrical contacts.

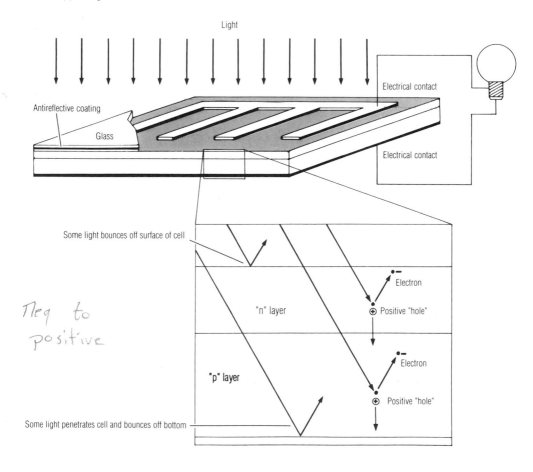

Light

Electrical contact

Antireflective coating

Glass

Electrical contact

Some light bounces off surface of cell

Electron

"n" layer

Positive "hole"

Πeq to positive

Electron

"p" layer

Positive "hole"

Some light penetrates cell and bounces off bottom

trated) light. Flat-plate modules ready for commercial production have attained an efficiency of over 15 percent (DOE 1989c).

The main disadvantage of single-crystal silicon cells is their comparatively high cost. The cells must be fairly thick—100-300 micrometers, or millionths of a meter—to absorb sunlight and so require a lot of material, and their production process, in which silicon crystals are grown and then cut into wafers, is wasteful and slow. An approach to making single-crystal silicon more economical is to mount the cells on concentrating collectors (reflectors or lenses), greatly reducing the cell area required, though increasing collector costs. Placing the cells under concentrated light also raises their efficiency. The champion silicon concentrating cell has demonstrated an efficiency of 28.2 percent under light concentrated 100 times (DOE 1989c).

An alternative to single-crystal silicon is polycrystalline silicon, which has somewhat lower efficiency but costs less to manufacture. In one production method, molten silicon is drawn into sheets (called ribbons), while in another, it is cast as ingots and sliced into wafers. The highest efficiencies measured for individual cells are in the range of 15-17 percent (normal light). However, in flat-plate modules the efficiency drop is partially made up by better use of space, since polycrystalline cells are square.

The most radical departure from the high-efficiency, high-cost single-crystal cells is the relatively new thin films made from amorphous silicon (a-Si). Since amorphous silicon absorbs light easily, the films can be very thin—just one or two micrometers thick. Moreover, thin films can be deposited on various substrates, such as glass or metal, in a way that is amenable to mass production. Their principal drawback is relatively low efficiency. The laboratory record for small cells is 12 percent, and today's commercial modules have an efficiency of less than 7 percent. Also, the efficiency of amorphous-silicon cells declines somewhat after exposure to sunlight. The degradation, once as high as 50 percent, has been reduced to about 10 percent in single cells and 15 percent in modules (Hubbard 1989b). The question remains whether amorphous-silicon thin films will prove as durable as crystalline-silicon cells, which are expected to last 20 years or more. The Department of Energy has set a goal of 30-year operation for both types of materials (DOE 1987).

A variety of other materials besides silicon have begun to emerge as contenders in the photovoltaic market. The most efficient of all cell materials is gallium arsenide (GaAs). Efficiencies of 24.2 percent under normal light and 29.2 percent under concentrated light have been measured with gallium-arsenide cells in the laboratory. The cells are expensive and difficult to manufacture in large quantities, however. Another important advance has been the development of copper indium diselenide ($CuInSe_2$) and cadmium telluride (CdTe) thin films with efficiencies comparable to or greater than amorphous-silicon thin films. Initial problems with the materials appear to have been solved, opening the way for potentially rapid improvements in performance and reductions in cost. They may be even more suitable for mass production than amorphous silicon (Hubbard 1989b).

Table 4. Current and potential efficiencies of small photovoltaic cells. Source: Department of Energy (1989c).

Type	Small-Cell Efficiency (%)	
	Current	Potential
Flat Plate		25-30
Single-crystal Si	22.8	
GaAs	24.2	
Concentrator		30-40
Si	28.2	
GaAs	29.2	
2-junction GaAs/Si	31.0	
2-junction GaAs	34.2*	
Thin Film		15-20
GaAs	22.4	
Polycrystalline Si Ribbon	14.2	
a-Si	12.0	
$CuInSe_2$	12.9	
CdTe	11.0	
2-junction a-Si/$CuInSe_2$	14.6	
3-junction a-Si	13.3	

(handwritten annotations: "Gallium Arsenide" next to GaAs; "Copper Indium Diselenide" next to $CuInSe_2$; "Cadmium" next to CdTe)

* Preliminary result

To raise the efficiency of photovoltaic cells, two or three materials can be stacked in a multilayer cell to tap a broader spectrum of sunlight. One company, Boeing, has reported achieving a remarkable efficiency of 37 percent with a multilayer gallium-arsenide cell under concentrated light, although preliminary results of government tests indicate the actual efficiency of the cell is about 34.2 percent (Boes 1989). The best performance previously recorded for a photovoltaic cell was 31 percent with a tandem silicon/gallium-arsenide cell under concentrated light (DOE 1989c). The maximum theoretical efficiency of photovoltaic cells is limited by physics. Nevertheless, there is still considerable room for further improvement in laboratory-cell efficiencies, and commercial cells remain 5-10 years behind the state of the art. (See Table 4 for a summary of current and theoretical laboratory-cell efficiencies.)

Along with the improving performance of photovoltaic cells has come decreasing cost. In 1976, the price of photovoltaic modules averaged $44,000 per peak kilowatt of capacity (1986 dollars), but by 1986 the price had dropped to just over $5,000 per peak kilowatt, and the current price is estimated to be $4,000-$5,000 per peak kilowatt[3] (IEA 1987, Shea 1988). Other components—power conditioners, regulators, direct-to-alternating current converters, and controls—will approximately double the cost of installed systems. Although progress has been impressive, the price is not as low as was once predicted. Ten years ago, the Department of Energy set cost goals of $3,800 per kilowatt in 1982, $1,000 per kilowatt in 1986, and $200-$600 per kilowatt in 1990 (Kendall and Nadis 1980, converted to 1986 dollars). A decrease of more than 80 percent in government funding for research is one factor behind the failure to meet these early goals, though unexpected technical difficulties (such as degradation) and the lack of a secure domestic market have also played a part.

As costs have fallen new markets for photovoltaic cells have opened. In 1976 annual world shipments of photovoltaic modules amounted to 0.5 MW, and by 1983 shipments had grown to 20 MW. After 1983, photovoltaic sales held steady for a few years, but more recently they have begun rising again, reaching about 33 MW in 1988 (DOE 1989c). The photovoltaic industry has undergone profound changes. Single-crystal silicon cells initially dominated sales, but gradually their share has declined as new materials—mainly polycrystalline and amorphous silicon—have entered the market. At the same time, the US industry's once-overwhelming share of the world market has fallen to between one-third and one-half. The main cause of this change is the rising importance of the Japanese industry, which has developed amorphous-silicon cells for consumer products. The US industry has traditionally concentrated on crystalline silicon for power applications, although several manufacturers (e.g., Chronar and Solarex) are producing thin films.

Currently the largest photovoltaic market—about 15 MW per year—is powering equipment and communities far from utility grids (DOE 1989c). (Consumer products, such as solar-powered calculators, are the second-largest market, but they have no significance for energy supply.) Many remote communications systems and navigation aids now rely on photovoltaic systems to provide power, eliminating the need to recharge batteries or maintain and refuel diesel-electric generators. In the United States and Europe, photovoltaic systems are frequently installed in vacation homes (15,000 in the United States, according to NAHB, 1988), with typical units of a few hundred watts capacity costing $2,000-$5,000 (IEA 1987, Real Goods News 1989). The total potential market for such systems, according to one estimate, is a few tens of gigawatts, representing tens of billions of dollars at current prices (IEA 1987).

Providing power for villages in less-developed countries is a fast-growing market with very large potential. The United Nations estimates that over two million villages worldwide are without electric power for water supply, refrigeration, lighting, and other basic needs. The cost of extending utility grids to such areas is often prohibitive ($23,000-$46,000 per kilometer, according to Shea, 1988). Photovoltaic systems could meet much of that potential need. They are now competitive with diesel-electric generators (two million of which are sold yearly) for small-scale applications of less than a few kilowatts peak capacity (DOE 1987). To what extent less-developed countries will have the capital to invest in photovoltaic systems (or any other energy technologies) is unclear, however.

Photovoltaic-Dish System Cost vs. Annual Production

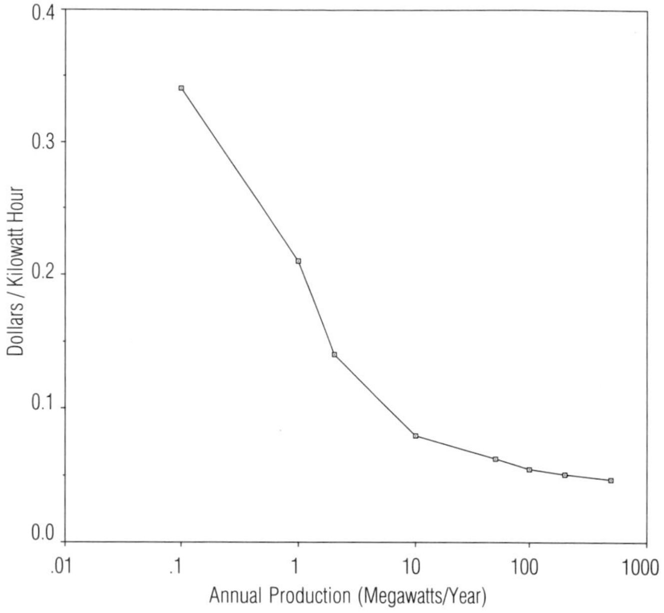

Figure 13. The estimated cost of electricity produced by photovoltaic-dish systems as a function of the annual rate of factory production. Source: Davenport (1989).

In the United States and other industrialized countries, the most important future market for photovoltaic systems will probably be larger peak-power stations and smaller grid-connected homes and businesses. For the moment, this market is still out of reach. The cost of electricity generated by photovoltaic systems is in the range of 25-35¢/kWh, depending on the size of the system, or over twice the cost of electricity generated by Luz solar-thermal electric plants (DOE 1989c). However, the Department of Energy (1987) projects that costs will drop to 12¢/kWh by the early 1990s and possibly to 6¢/kWh by the turn of the century. Assuming these cost goals are achieved, photovoltaic systems could begin to compete in the electric-utility peak-power market in the mid-to-late 1990s (DOE 1987). The two most important requirements for achieving lower costs will be producing commercial modules with the high efficiencies of the laboratory and manufacturing photovoltaic systems in much larger quantities. According to an analysis by Science Applications International Corporation, the cost of electricity produced by photovoltaic systems with parabolic-dish concentrators could be reduced below 5¢/kWh with the manufacture of more than 100 MW of peak capacity annually (Davenport 1989). (See Figure 13.)

Greater experience and familiarity with photovoltaic systems will be essential for electric utilities to have confidence in this technology. Over a dozen large demonstration power plants (greater than 100 kilowatts peak capacity) and numerous small ones have been built worldwide since 1981 (IEA 1987). The largest is a 6.4 MW plant using single-crystal silicon cells built in 1983 by ARCO Solar for Pacific Gas and Electric in Carissa Plains, California. Experience with such plants has shown that, in general, photovoltaic systems are well-suited to utility applications: they produce consistent power that is well matched to daytime load profiles, their performance is well understood and predictable, failure rates of modules are low (0.5-1 percent or less per year), and operations and maintenance costs are minimal (about 0.5¢/kWh or less) (SNL 1987a, 1987b).

While the number of planned demonstration projects has declined in recent years, several are still in the works. The most important is Photovoltaics for Utility-Scale Applications (PVUSA), a demonstration plant jointly funded by the federal government and electric utilities (led by Pacific Gas and Electric) that will compare the performance and cost of, and provide experience with, systems from several different manufacturers. A number of manufacturers have also embarked on demonstration projects with individual utilites.

At the other end of the power spectrum, photovoltaic systems can be installed in homes, communities, or businesses connected to electric-utility grids. Several experimental residential projects, such as one in Gardner, Massachusetts, that serves 35 homes, have been established around the country (NAHB 1988). The advantage of this approach is that average electricity prices are higher for end-users than at the busbar (where electricity enters the grid), thus raising the incentive for end-users to produce their own electricity. On the other hand, small-scale photovoltaic systems are more expensive per kilowatt than large-scale systems, although with standardization of design economies of scale are still possible. In addition, electric utilities rarely charge residential consumers time-of-day rates to reflect the true cost of electricity during peak periods, and it is on these higher rates that photovoltaic systems will depend for their initial viability. Thus rate structures will have to be changed for small-scale, on-site photovoltaic systems to be a success. Nevertheless, the long-term prospects for such applications appear bright, since photovoltaic systems are simple, require little maintenance, and pose no unmanageable safety hazards.

Energy Storage

Sunlight is variable and unavailable at night, so energy-storage systems will eventually be necessary if solar energy is to be widely exploited. Pumped hydroelectric storage (the transfer of water between low and high reservoirs) is already in wide use and could become even more important in the future (see the hydroelectric-power chapter). Other electrical-storage systems, such as advanced batteries and compressed air, are not yet cost effective but appear promising (OTA 1985, DOE 1989d). Thermal storage usually involves heating masses of water or some dense material (rocks or bricks, for example) and is an integral part of solar-building designs. With a large, insulated mass underground, it is possible to store energy for months, so that heat collected in summer could be used in winter. Experiments conducted in Sweden suggest that the cost of such a seasonal storage system for a small community could be about $11/GJ, less expensive than electric heat but more expensive than natural gas and fuel oil (IEA 1987). The use of materials now under development that change phase or chemical composition with changes in temperature could reduce the mass of storage material required and provide a built-in mechanism of temperature regulation.

An intriguing possibility is that energy could be stored in the form of hydrogen produced by the splitting of water molecules with electricity generated from renewable sources. Hydrogen is a clean-burning gas that can be used to generate electricity, provide heat, and even power automobiles. While hydrogen is currently expensive, some analysts believe that with large-scale development of renewable electricity sources, particularly photovoltaic cells, it could become a competitive alternative to fossil fuels (Ogden and Williams 1989).

Although the need for energy storage must be taken into account, its importance should not be exaggerated. Energy storage will not be required for many near-term applications, particularly electricity generation. Electric utilities already have considerable reserve capacity—typically 20 percent of peak demand—so that the unexpected shutdown of a few plants will not cause power shortages, and additional reserves will not be required until solar-generated electricity constitutes a significant fraction of the total electricity supply. (See the wind-power chapter for more details.) In addition, variations in insolation are often closely matched to variations in energy demand, thus reducing the need for storage. Peak demand for electricity and industrial thermal energy often occurs during the day. The need for storage or backup could be further reduced by integrating several different renewable and fossil energy sources in a single energy-supply system. In particular, using natural gas as a backup fuel source could be a cost-effective substitute for storage in applications ranging from electricity generation to residential water and space heating.

Conclusions and Near-Term Prospects

A wide variety of solar-energy technologies are at varying stages of development. While none is yet providing a significant amount of energy, several are or soon could be in a position to do so. Among solar-building technologies, passive-solar heating is the most cost effective. Well-designed solar homes cost little more to build than conventional homes but require much less fossil energy for heating, cooling, and lighting. However, passive-solar designs have been slow to be accepted in the housing industry and have made almost no impact on commercial buildings. The history of active-solar building technologies is likewise mixed. The US residential solar-collector market, after enjoying a brief boom in the early 1980s, shrank by three-fourths in the late 1980s. In the present economic climate, space- and water-heating systems are too expensive to be competitive in most parts of the country, except where sunshine is abundant and electricity is the competing fuel. Their costs could fall sharply, however, if the market demand were large enough to justify mass production of systems.

Solar-thermal electric systems are making gradual headway in the electric-utility market. Several plants have been built by one company, Luz Industries, in Southern California. New and larger plants soon to come on line are expected to be competitive, or nearly so, with conventional technologies for generating peak power. The costs of electricity produced by parabolic-trough, parabolic-dish, and central-receiver systems could fall as low as 6-8¢/kWh by the mid-to-late 1990s, and faster progress is possible if market demand grows quickly. (In areas with less abundant sunshine than the Southwest, costs will be higher.)

Although solar-thermal systems are well suited to supplying industrial process heat, this application has not been significantly developed. One reason for the lack of industry interest is the absence of regulations, similar to those governing utilities, favoring alternative energy sources. One possible near-term application of concentrating solar collectors is the breakdown of hazardous chemicals. A recent Department of Energy experiment showed that dioxins could be almost entirely destroyed by a combination of heat and concentrated sunlight. Other solar photochemical applications, such as photoenhanced conversion of inexpensive chemical feedstocks into liquid fuels, are also being explored (DOE 1989b).

In the past decade, the cost of electricity generated by photovoltaic cells has fallen 90 percent. Already economical in areas remote from utility grids, photovoltaic systems could begin to compete in the electric-utility market by the mid-to-late 1990s. Some experts believe that by the end of the 1990s, photovoltaic electricity could cost as little as 6¢/kWh (Bath 1989). It is not yet clear which among the various technologies—thin films, crystalline-silicon flat-plate modules, or concentrator systems—will be the most cost effective for power applications.

The photovoltaic industry has been fortunate that several niche markets for photovoltaic cells—in particular, consumer products and remote power—have given companies the opportunity to expand production and lower costs without having to meet the more demanding economics of the utility market. The industry is not yet profitable, however. American manufacturers are for the most part subsidized by much larger companies (typically oil companies), some of which are showing signs of losing patience with continued losses (Wald 1989). Photovoltaic-cell manufacturers rely heavily on government funding for research and development, and greater support will be essential to ensuring a viable industry in the United States in the 1990s.

With no change in federal policies, solar energy will probably play only a small part in the US energy picture for the next decade or two. According to a Department of Energy projection, solar technologies will supply the equivalent of 0.5 percent of projected US primary-energy demand in 2000 and 1.7 percent in 2010, compared to 0.05-0.075 percent today. One-third to one-half of this energy will be supplied by photovoltaic systems and the rest by large-scale and small-scale solar-thermal systems (DOE 1988a).

The commercialization of solar technologies could be greatly accelerated, however, with additional market incentives, such as solar-energy tax credits; new regulations requiring passive- and active-solar technologies to be included in new building construction; and greater federal funding of solar-energy research and development. Sklar (1989) estimates that under an "enhanced" scenario, solar technologies could supply more than 6 EJ (8 percent of current US energy demand) by 2000, with solar-thermal energy, solar buildings, and photovoltaic cells making approximately equal contributions. This estimate may be optimistic; in our opinion, a 1-2 EJ solar-energy contribution by 2000 seems a more feasible goal. This goal could be met if, for example, passive-solar designs were incorporated into one-quarter of all new housing construction between now and 2000, solar water and space heating were installed in one-tenth of existing houses, 10,000 MW of solar-thermal and photovoltaic electrical-generation capacity were installed, and solar-thermal systems met 1-2 percent of industrial heating needs.

Notes

1. This figure assumes an average heat production of 54 MJ/day for a 5-m² system and a cost of electricity of 10¢/kWh ($27/GJ).

2. This figure assumes a 15 percent fixed-charge rate and an average daily heat production of 6.8-8.6 MJ/m².

3. Peak watt refers to the power generated at maximum solar incidence of about 1,000 watts/m². Thus, a 1-m², 10 percent efficient module would have a peak capacity of about 100 watts and today would cost $400-$500 (not including installation and balance-of-system costs).

Wind Energy

Wind energy has been used for many centuries to pump water, thresh grain, and propel ships. In the early part of this century, windmills for pumping water were common in rural areas of the United States, and over six million are estimated to have been produced (Metz and Hammond 1978). The extension of electric-utility service to most parts of the country in the 1930s and 1940s caused many of these reliable and durable machines to be abandoned.

Rising energy prices brought about a revival of wind power in the United States in the 1970s and 1980s. Between 1981 and 1988, 15,658 wind turbines with a total peak capacity of 1,370 megawatts (MW) were installed in California, where the vast majority of US and world wind-power development has taken place. In 1988, the turbines, located at three main sites, the Altamont, Tehachapi, and San Gorgonio Passes, generated 1.827 billion kilowatt-hours of electricity, about 1 percent of California's demand and as much as a city the size of San Francisco consumes in a year (Gipe 1989, Lynette 1989a).

Wind power has the potential to supply a large fraction—probably at least 20 percent—of US electricity demand at an economical price. It is easily the least expensive of the emerging renewable electricity sources. Existing wind turbines are producing electricity at a cost of about 7-9¢/kWh. Within the next few years, the cost could be reduced to 4-6¢/kWh at windy sites and somewhat more at less windy sites, making wind power competitive with conventional electrical-generation technologies in most utility markets.

However, wind-power development has greatly slowed since 1985, when renewable-energy tax credits were eliminated and fossil-fuel prices began falling precipitously. The amount of wind-power capacity installed annually in California has fallen over 80 percent, from 398 MW in 1985 to 67 MW in 1988 (see Figure 14). Federal support for wind-energy research and development has been cut 85 percent in eight years, from $60.1 million in fiscal 1980 to $8.8 million in fiscal 1989 (Sissine 1989). Not surprisingly, many US wind-power developers and manufacturers have gone out of business, and the world wind-turbine market is increasingly dominated by foreign, particularly Danish, manufacturers. Unless there is a dramatic shift in federal and state policies toward renewable energy sources, wind power is unlikely to see a resurgence in the United States for several years.

The Wind Resource

The amount of wind energy theoretically available in the United States is estimated to be 3,000 exajoules (EJ) per year, or 40 times current annual energy consumption of 84 EJ (NSF 1972). Only a small fraction of this resource could be exploited because of constraints on available land and the efficiency of energy extraction. But while estimates of the extractable resource vary, and further research is necessary to determine its true size, the consensus appears to be that it is very large. Based on a review of other studies, Kendall and Nadis (1980) conclude that wind power could supply 1-4 trillion kWh annually (40-150 percent of current electricity demand) by making use of suitable land not currently used for other purposes. The wind resource is concentrated along the Pacific and Atlantic coasts and in the Great Plains (see Figure 15).

Technologies and Economics

Economics will play a central role in determining how much of the wind resource will actually be developed. The amount of energy a wind turbine can produce increases as the cube of the wind velocity. Thus, high-wind sites (those with average wind speeds of more than 15 mph) are more economical than moderate-wind (12-15 mph) and low-wind (less than 12 mph) sites and should be exploited first, as indeed they have been in California. However, suitable high-wind sites are relatively rare. One study estimated that enough wind-power capacity could be installed in the United States to supply about 40 percent of electricity demand; but of that amount, only 4 percent would be installed at high-wind sites while 55 percent would be installed at low-wind sites (GE 1977). Realizing wind power's full potential will thus depend on lowering its costs at least to the point where moderate-wind sites are economical. This point has not yet been reached, though steady progress is being made.

Wind turbines are basically simple machines consisting of blades, rotor, transmission, electrical generator, and control system, all mounted on a tower (see Figure 16). The vast majority of wind turbines now in operation are intermediate in size (approximately 50-300 kilowatts peak capacity[1] and 15-30 meters rotor diame-

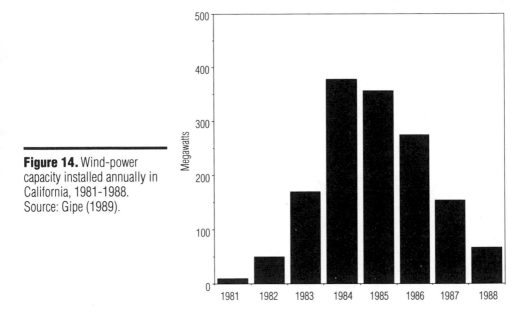

Figure 14. Wind-power capacity installed annually in California, 1981-1988. Source: Gipe (1989).

ter) with a two- or three-bladed rotor mounted on a horizontal axis. In most horizontal-axis machines, the rotor is downwind of the tower and free to align itself with the wind, although in some the rotor alignment, or yaw, is actively controlled to keep it upwind of the tower. A small number of wind turbines are of the hoop-shaped Darrieus design, which is mounted on a vertical axis and is insensitive to wind direction. Much smaller and much larger wind turbines have also been developed but have not been as commercially successful as intermediate-size machines.

Although basic wind-turbine designs have changed little, a number of improvements over the years have greatly increased their performance. In the early 1980s, wind-turbine reliability was a serious problem as some components proved susceptible to fatigue, vibration-failure, and mechanical breakdowns. With more rugged designs, better choice of materials (e.g., fiberglass instead of aluminum blades), and more careful attention to maintenance, wind-turbine availability has risen to over 80 percent, and mature systems are now operating 95-98 percent of the time (IEA 1987). The overall efficiency of wind turbines (energy extracted per square meter of swept rotor area) has also improved through a combination of better siting, more efficient designs, and higher reliability. For example, the efficiency of 55-kilowatt Danish wind turbines has increased over 50 percent, from 395 kWh/m^2/year in 1981 to 641 kWh/m^2/year in 1986 (Gipe 1989).

Figure 15. Mean annual wind-power density (watts/ square meter) over the United States. (Over mountainous regions [shaded areas], the estimates are the lower limits expected for exposed summits and ridges.) Source: Kendall and Nadis (1980).

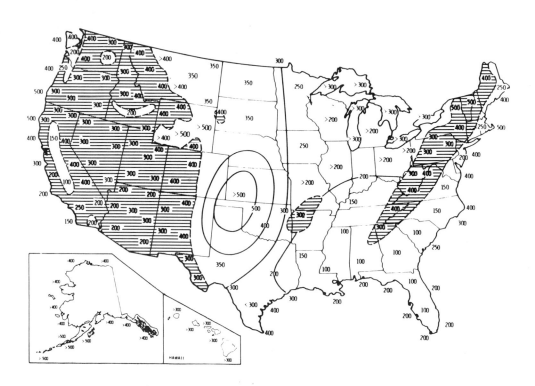

Further efficiency improvements are possible and probably necessary if wind power generated at moderate-wind sites is to be competitive with conventionally generated electricity. Options for raising performance include adjusting blade-pitch angles and rotor speeds to suit lower wind speeds, improving the design of blade airfoils, developing variable-speed/constant-frequency electric generators to allow wind-turbine rotors to change speed with the wind, and modifying control strategies to prevent unnecessary shutdowns in high winds (IEA 1987, DOE 1988b). An important task is to develop advanced designs while still maintaining high reliability. US wind-turbine manufacturers have traditionally favored innovative technologies for achieving higher performance, while Danish manufacturers have favored sturdiness and dependability (Dodge and Thresher 1989). Both qualities will be necessary in future designs.

As designs and performance have improved, the cost of wind power has declined dramatically. The levelized cost of electricity generated by modern wind turbines is now estimated to be 7-9¢/kWh, down from over 25¢/kWh in 1981, while the average price per installed kilowatt of intermediate-size wind turbines declined from $1,300-$2,000 in 1981 to $950-$1,100 in 1988 (AWEA 1988a, Lynette 1989b). One advantage of wind power is that operations-and-maintenance (O&M) costs can be quite low. O&M costs at operating wind farms range from 0.8 to 2¢/kWh and average 1.2¢/kWh (AWEA 1988b). By comparison, the operations, maintenance, and fuel costs of coal and nuclear plants in 1986 were 2.2 and 2.0¢/kWh, respectively (EIA 1988b).

Despite its relatively low cost, wind power can make only limited headway in the present electric-utility market. Initial wind-power development in California took advantage of long-term, fixed-price contracts (Standard Offer 4) offered by electric utilities in the early 1980s at rates upwards of 6¢/kWh. Today, in areas with good wind resources, utilities generally have avoided costs of 3-4¢/kWh, too low to encourage further wind-power development (Lynette 1989a). Avoided costs are much lower than anticipated a few years ago partly because fossil-fuel prices are lower and partly because most utilities have considerable excess generating capacity, allowing them to operate their more efficient plants most of the time. (The avoided cost is the price utilities are required to pay independent power producers under PURPA regulations; for more details, see Chapter 7.)

However, wind power's competitive position is likely to improve by the mid-1990s. By then, its cost is expected to be about 6¢/kWh at windy sites (CEC 1988a, Lynette 1989a). Utility avoided costs are likely to be at or above this level as excess capacity is absorbed and fossil-fuel prices rise. With improved performance and mass production of wind turbines, it should be possible to provide wind power at a cost of 4¢/kWh at high-wind sites, well below projected costs of conventional electrical-generation technologies (OTA 1985, DOE 1988b). The cost would be somewhat higher (perhaps 6-8¢/kWh) at moderate-wind sites.

Residential and commercial markets for wind turbines are likely to grow much more slowly than the utility market. The smaller wind turbines that would be required are presently more expensive per installed kilowatt than intermediate-size turbines, and O&M costs are likely to be higher as well, until an industry to service and repair the machines is established.

Principal Components of a Wind Turbine

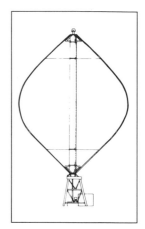

Figure 16. In most wind turbines, the rotor is mounted on a horizontal axis (immediate right and below), but in some it is on a vertical axis (far right).

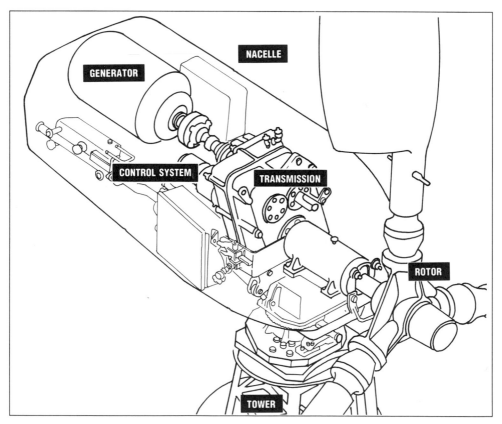

GENERATOR

NACELLE

CONTROL SYSTEM

TRANSMISSION

ROTOR

TOWER

Energy Storage

Because wind speed fluctuates over time, making extensive use of wind power, like solar energy, will require energy storage. (For a discussion of energy-storage options, see the solar-energy chapter.) Coupled with a modest amount of storage capacity, wind energy can be remarkably reliable. Justus (1976) estimated that an interconnected network of wind farms could supply electricity with 95 percent reliability if 24 to 48 hours of storage capacity were built into the system, while Sorensen (1976) concluded that the reliability of a single wind turbine with 10 to 24 hours' storage capacity would be higher than 60 percent and comparable to the reliability of a nuclear power plant.

The problem of the intermittency of wind power is often overstated, however. Although winds are inherently variable, they often follow daily and seasonal patterns that are surprisingly predictable, as sailors following trade winds know well. Furthermore, the reliability of wind power can be increased by linking a number of widely separated wind farms in a single grid, so that a wind drop in one area would often be offset by a wind increase in another. Intermittency will not be a problem in the near term in any event, as electric utilities normally have a substantial reserve capacity to handle unexpected supply breakdowns. This reserve should suffice until wind power constitutes a substantial fraction of the total electricity supply. Estimates of the allowable wind fraction range from a few percent for a grid carrying a single wind farm to 20 percent or more for one carrying a number of widely separated farms (Kendall and Nadis 1980, Grubb 1988).

Environmental Impacts

Wind turbines are among the most environmentally benign sources of electricity. They produce no air or water pollution, and even large wind farms do not affect local wind and weather patterns in any significant way. Wind turbines create noise and are visually objectionable to some people, and have been known to kill birds, but these problems should not be important if wind farms are located away from population centers and scenic or otherwise sensitive areas.

Land use is often cited as a significant constraint on wind-power development. A General Electric study estimated that over 100,000 square miles of land—an area 40 percent the size of Texas—would be needed to produce 40 percent of present US electricity demand (GE 1977). The capacity per square mile is limited by the need to space wind turbines far enough apart to avoid wake interference and turbulence. The GE study assumed an average spacing of 15 rotor diameters, resulting in a peak-power density of about 2-4.5 MW per square mile.

In practice, however, wind turbines in California are much more densely spaced than the GE study assumed (AWEA 1989b). Moreover, the wind turbines themselves occupy only a small fraction of the land area, and they do not exclude other land uses, particularly farming and ranching. In California, the presence of wind turbines on farmland has increased land values and revenues. Perhaps the greatest potential for wind-power development is in the Great Plains, where vast stretches of farmland could support hundreds of thousands of wind turbines. In other areas, however, the construction of wind turbines could require clearing trees and cutting roads in forested areas, a prospect that is sure to generate controversy and public opposition, especially since, as Bain (1988) wrote, "Standards in effect at most sites [in the Pacific Northwest] are inadequate to protect the environment."

Conclusions and Near-Term Prospects

Wind power is probably in the best position of any of the renewable electrical-generation technologies to provide a significant portion of US electricity supply in the near future. The cost of wind power at high-wind sites is now competitive, or nearly so, with electricity generated from conventional coal-fired plants and other fossil-fuel sources. There is every likelihood that wind power's cost will continue to decline into the 1990s, making moderate-wind sites also economical. Enough wind turbines could probably be built on available land to meet over 40 percent of current US electricity demand. Since about half of those sites have moderate and high winds, it is reasonable to estimate that wind turbines could meet at least 20 percent of demand at an economical price in the next 20-30 years.

The near-term growth of wind power will be critically dependent on federal and state policies and market incentives. In the present electric-utility market and without additional incentives it is likely that wind-power development, already curtailed since 1985, will proceed slowly for the next several years. According to one projection, wind-power capacity in the United States will probably grow to 2,200-6,500 MW in 2000—at best a fourfold increase—under "business-as-usual." However, under an "enhanced" scenario involving more intensive research and development and the inclusion of environmental costs in the price of fossil-fuel sources (see Chapter 7), it is estimated that wind-power capacity could reach 6,100-32,300 MW by 2000, and electricity production could reach 7.84-41.54 billion kWh (0.3-1.5 percent of current demand) (Lynette 1989a). The lower part of this range is, if anything, conservative, and it is conceivable that considerably more than 30,000 MW of wind-power capacity could be in place by that time.

Notes

1. The peak capacity of a wind turbine is an often misleading measure of its electrical output, as it does not usually reflect realistic operating conditions and is not based on a standard wind speed.

Biomass Energy

Before the use of coal became widespread, biomass—principally firewood—was the most important energy source in the United States. For much of this century, its importance steadily declined, reaching a low in about 1960. Since then, however, the trend has reversed, and the consumption of wood and other biomass—plant matter—has grown to about 4-5 percent of US primary-energy consumption. The trend has been led by the forest-products industries, which use inexpensive wood and pulp waste to meet a large part of their energy needs. Since the 1970s, home wood burning has also become popular. Other sources of biomass energy now in use include municipal wastes and methane produced in landfills, as well as fuel ethanol produced from corn and other grain surpluses.

Biomass could play a much more important role in our energy future. Although the total resource is difficult to determine, approximately 6-8 exajoules per year (7-9.5 percent of current primary-energy demand) probably could be supplied by forestry, agricultural, and municipal wastes, and a comparable or larger fraction could be derived from trees and plants grown specifically for energy. A key advantage of biomass is that it can be converted into liquid and gaseous fuels, such as ethanol and methane, which can be substituted for petroleum products and natural gas. Biomass fuels (biofuels) could potentially replace 30-90 percent of US gasoline and diesel-fuel consumption, assuming that fuel-economy improvements constrain demand to current levels. Though still too expensive to succeed commercially in major markets, biofuels—particularly ethanol derived from wood and plants and gases derived from wastes—could become competitive in the 1990s.

The Biomass Resource

Biomass energy can be derived either from wastes or from trees and plants grown on energy farms. Calculating the energy potential of each source involves first estimating the amount of biomass that could be produced and collected at an economical price, then multiplying by the energy content of the biomass and, if the biomass is converted to a liquid or gaseous fuel rather than burned directly, by the conversion efficiency. Uncertainties in all three factors lead to significant discrepancies among various estimates of the total biomass resource. The estimates presented below are intended to give an idea of the general magnitude of the resource, not to set absolute limits.

Wastes

Wastes include wood and pulp from forest-products industries (logging, lumber, and paper), agricultural residues such as plant stalks and leaves left after harvesting and processing, animal manure, food and crop surpluses, and the organic components of municipal wastes. Resources of wood and wood-product wastes are probably the largest. The forest-products industries produce 150-300 million dry tons of waste each year with an energy content (before conversion to fuels) of roughly 2.25-4.5 exajoules (EJ), based on an average energy content of 15 gigajoules (GJ) per dry ton (Kendall and Nadis 1980, Erickson 1986c, IEA 1987). Of these wastes, close to 100 million tons are already being consumed for energy. Much of the remaining wastes consists of dispersed logging residues, which may be difficult to collect and expensive to transport. Thus, considerably less than 2.5-4.5 EJ may actually be economically recoverable.

The amount of energy available in agricultural wastes (plant stalks, leaves, etc.) is theoretically large, although in practice only a small fraction could probably be recovered and used because the residues are widely dispersed. The extractable energy potential is commonly estimated to be about 1 EJ, including energy losses from conversion to liquid or gaseous fuels (DOE 1988b, IEA 1987). Crop surpluses, another form of agricultural wastes, could supply an estimated 2-2.5 billion gallons of ethanol each year (Finneran 1986b), representing about 0.16-0.2 EJ of energy or 1.2-1.5 percent of current US gasoline consumption.

Cities and towns generate about 240 million tons of municipal waste each year, much of which consists of combustible organic materials (paper, yard waste, food waste, wood, containers, and packaging) with an estimated energy content of about 1.8 EJ. In addition, methane produced by landfills and sewage could be captured and burned, producing an estimated 0.3 EJ, and industrial solid wastes could contribute about another 0.2 EJ (Klass 1988, Finneran 1986a).

Table 5. Estimated recoverable energy potential of biomass wastes and energy farms, in exajoules per year.

Source	Total	As Fuels*
Wastes	5.75-8.0	1.7-3.7 (L&G)
Forestry	2.25-4.5	0-2.0 (L)
Agricultural	1.0	1.0 (G)
Surpluses	0.2	0.2 (L)
Municipal and Other	2.3	0.5 (G)
Energy Farms	6.75-18.5	4.37-10.25 (L)
Woody and Herbaceous	4.0-15.0	2.0-7.5 (L)
Plant Oils	2.0	2.0
Conventional Forestry	0.75-1.5	0.37-0.75
Total	12.5-26.5	6.07-13.95 (L&G)
% US Energy Consumption	15-32%	
% Motor Vehicle Consumption (L)		30-90%

* Fuel potential calculated assuming 50% conversion efficiency, except for fuels (e.g., methane from landfills) produced directly. Substitution for gasoline assumes 15 percent increase in fuel economy (see Note 2 at end of chapter). L=liquid, G=gaseous fuels.

Table 6. Comparison of estimates of the maximum recoverable biomass resource, in exajoules per year. (For the purpose of comparison, exajoules and quads are assumed equal.)

Resource	UCS	K&N	DOE	Klass	Lynd*
Wastes	5.75-8.0	2-10			5.0
Farms	6.75-18.5	2-10			2.9-18.6
Total	12.5-26.5	4-20	17.0	14.6	7.9-23.6

* As ethanol fuel.
Sources: UCS, this report; Kendall and Nadis (1980); DOE (1988d), Klass (1988); Lynd (1989).

Energy Farms

The main constraint on growing plants and trees specifically for energy is the large amount of land that would be needed because of the relatively low efficiency of photosynthesis. As a rough rule of thumb, 20 million acres of cropland could produce, at present, 100 million tons per year of dry biomass with an energy content of 1.5 EJ if burned directly or 0.75 EJ if converted to liquid fuels.[1] To produce enough liquid fuels to replace all of the gasoline and diesel fuel consumed annually in the United States (about 18 EJ) would thus require planting almost 500 million acres, or more land than the United States now devotes to agriculture.

Nevertheless, there is considerable room to grow energy plants and trees on land not presently used for agriculture. Between 25 and 75 million acres of farmland are idle in any given year as part of the Conservation Reserve Program and other federal farm programs (Lynd 1989). If one includes potential cropland (including pastureland, rangeland, and forest, as defined by the US Department of Agriculture) the total rises to over 200 million acres. Not all of this land would actually be available for use. But assuming that 50-200 million acres (2-8 percent of US land area) could be planted, then about 4-15 EJ of direct energy, or 2-7.5 EJ of biofuels, could be produced each year. Growing energy crops on a commercial scale will require fast-growing strains of plants and trees that use less water and fertilizer than typical food crops and can be readily converted to usable fuels. Research is focusing on short-rotation hardwood trees, such as poplars and sycamores, as well as herbaceous plants such as switchgrass and sorghum (DOE 1988d).

Energy could also be extracted from existing commercial forests. According to one estimate, approximately 100 million acres of commercial forest (excluding that classified as potential cropland) could be adapted to energy production (Office of Technology Assessment data cited by Lynd, 1989). At current average timber growth rates of commercial forests (cited by Marland 1989), this forest area could produce about 0.75 EJ/year in direct combustion or 0.37 EJ/year as liquid fuels. Under more efficient and intensive management, the yield could be at least twice as great.

Oilseed plants (such as sunflowers and rapeseed) are another potential energy crop from which oils can be extracted and substituted for diesel fuel and fuel oil. For example, winter rapeseed, generally grown in northern climates, could be grown on farms in the South during the winter off-season. A more speculative source of fuel oil is aquatic plants, such as microalgae, which could be grown offshore, in marshes, or in farms fed by saline groundwater. The Department of Energy estimates that up to 2 EJ could be derived from both sources (DOE 1988b).

Summary

Forestry, agricultural, and municipal wastes, if efficiently collected and converted to energy, could probably supply up to 5.75-8 EJ per year or 7-9.5 percent of current US energy consumption. The potential of energy farms is more difficult to assess but is probably in the range of 6.75-18.5 exajoules or 8-22 percent of US energy consumption, with the lower part of the range reflecting the near-term potential and the upper part the long-term potential (see Table 5). These estimates are in line with other estimates shown in Table 6. Although the total resource of 12.5-26.5 EJ represents at most one-third of current US energy consumption, its importance as a potential source of transportation fuels should not be ignored. Biomass-derived liquid fuels could replace approximately 30-90 percent of the gasoline and diesel fuel currently used for transportation.

Technologies and Economics

How much of the biomass resource can be developed will depend both on the costs of biomass technologies as well as on their environmental impacts (discussed below). Klass (1988) estimates that biomass consumption in 1987 was about 3.11 EJ, although Rinebolt (1989a) estimates current consumption to be 4.0 EJ. More than 90 percent of this energy was derived from the direct combustion of wood and wood products (see Figure 17). A number of technologies for converting biomass into liquid and gaseous fuels have been developed, but they have had comparatively limited commercial success.

Direct Combustion

Almost any sort of biomass can be burned to produce heat, steam, and electricity. Direct combustion of wood and wood products in 1987 generated 2.87 EJ, according to Klass (1988). Of this, home wood burning accounted for about 30 percent, consumption by the forest-products industries 66 percent, and consumption by other industries and electric utilities for the rest. In 1987, about 5 percent of American households burned wood as their main source of heat (EIA 1989b). The lumber and paper industries satisfy about 50 percent and 70 percent of their energy needs, respectively, with wastes from wood and wood products (Erickson 1986a).

Some of the energy produced in combustion is used to generate electricity. In 1988, a total of about 5,154 MW of mostly wood-fired electrical-generation capacity was in operation, up from 250 MW in 1981 (Klass 1988, Rinebolt 1989a). The great majority of the plants are owned by independent power producers, many of them in the forest-products industries. Utilities themselves have shown only slight interest in wood-fired electrical-generation plants. Between 1983 and 1987, electric utilities built just four plants of 45 MW capacity each (Shea 1988).

Wood-fired electrical-generation plants are not very different in design from conventional coal-fired plants, and capital and operating costs are similar (CEC 1985, 1987). The relative cost of electricity produced is strongly dependent

Biomass Energy Consumption in 1987 and 2000

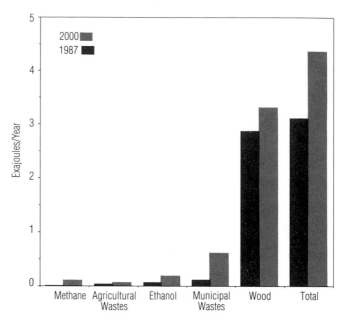

Figure 17. Consumption of biomass energy in 1987 and projected to 2000, assuming no change in market conditions or government policy. Source: Klass (1988).

on the cost of the feedstock. Wood waste produced at lumber and paper mills is virtually free, and in fact it can be more expensive to dispose of than to burn. Where wood waste is not readily available from mills, however, it must be collected from forest residues. In California, it is economical to collect and haul wood waste by truck over a distance of no more than 75-100 miles, thus effectively limiting the size of wood-fired electric-power plants to 25-50 MW (CEC 1987). The cost of electricity from utility-owned wood-fired plants completed in 1992 is projected to be 8-15¢/ kWh (CEC 1988a, converted to 1989 dollars). Concepts have been proposed that could make electricity generated from wood considerably less expensive. Whole-tree burning, for example, could permit larger plants to be constructed while reducing wood handling and processing costs (Ostlie 1989). In addition, gaseous fuels derived from biomass could drive high-efficiency, low-cost gas turbines (see the biochemical-conversion section below).

Municipal solid wastes (MSW) are also being burned for energy, most often because cities do not know what else to do with them. The costs of disposing of wastes have been rising in densely populated areas as nearby landfills and dumps have reached capacity and closed. As of 1988, a total of 105 waste-to-energy facilities were in operation, 84 of them burning raw waste and 21 burning refuse-derived fuels (RDF), a refined form of waste fuel from which most inorganic materials and moisture have been removed. Refuse-derived fuels have the advantages that they can be burned in conventional boilers and they contain fewer toxic pollutants. All told, waste-to-energy plants consume 6-10 percent of the wastes generated in the United States and produce an estimated 0.12 EJ each year (Klass 1988).

Political and economic pressures to convert municipal wastes to energy are sure to grow in the future as waste-disposal costs rise. However, it is unclear how many additional waste-to-energy facilities cities will be able to build. Several planned facilities have been blocked by local opposition sparked by such concerns as air pollution and toxic-ash disposal, the social stigma attached to living near a waste facility, and safety risks from increased truck traffic (CEC 1985). Twenty-five completed facilities have been permanently shut down for a variety of reasons, including environmental concerns and poor performance. Some cities are considering recycling most wastes as an alternative to burning them, and indeed this approach may have important environmental and economic advantages (EDF 1985, Newsday 1989). Recycling reduces the amount of energy needed to make products containing paper, glass, aluminum, and other recyclable materials.

A variety of processes for converting biomass to liquid and gaseous fuels have been developed (see Figure 18). The two main process categories are thermochemical conversion, which produces liquid methanol and syngas (a combustible mixture of carbon monoxide and hydrogen), and biochemical conversion, which produces ethanol and biogas (methane and carbon dioxide).

Thermochemical Conversion: Methanol and Syngas

Heating biomass in an oxygen-deficient or oxygen-free atmosphere breaks plant matter down into simpler solids, liquids, and gases. This principle has been known for centuries, and several techniques have been developed to produce different types of fuels as well as feedstocks for petrochemical industries. In general, the methods are less sensitive than biochemical processes to the type of feedstock. However, their products are sometimes troublesome to use, and the residues can be toxic and difficult to dispose of.

Of the three main thermochemical processes, gasification is the most widely commercialized (Erickson 1986b). Most gasification methods combine heat, steam, and air or oxygen to produce a mixture of carbon monoxide and hydrogen, known as syngas. In an earlier era this gas mixture, known as town gas, powered street lamps. Town gas was produced mainly from coal, but in fact biomass can be a better feedstock (IEA 1987). Using air in the process produces low-BTU gas (LBG), the most common type, with up to one-quarter the energy content of natural gas. Using pure oxygen yields medium-BTU gas (MBG) with up to one-half the energy content of natural gas. High-BTU gas can also be made from biomass, but the process is energy-intensive and far from economical (DOE 1988b).

A modest gasification industry was established in the early 1980s mainly to supply the forest-products industries, which use the gas generated from wastes to fire wood kilns and provide space heating. However, few gasification systems have been installed since fossil-fuel prices began falling in 1985, and all of the systems now in operation provide less than 0.006 EJ/year (Klass 1988). The potential market for syngas is large, since existing gas- or oil-fired boilers can be adapted to burn it. At presently low oil and gas prices, however, LBG and MBG are not economical except where there is a nearby source of low-cost waste and nearby demand for the energy produced. According to one estimate, the price of oil would have to rise above $40 per barrel ($7/GJ) for LBG systems to be broadly competitive (Erickson 1986b). The Department of Energy estimates that MBG currently costs about $8/GJ to produce but predicts that with further gains in production efficiency the cost could fall to $3.50/GJ—comparable to the current price of natural gas for industrial users—by 1992 (DOE 1988b).

Pyrolysis has traditionally been used to make charcoal from wood. It is a relatively simple process, requiring no steam or purified oxygen, and it produces a mixture of char, liquids (oils and methanol), and gases (methane, carbon monoxide, carbon dioxide), with the proportions determined by the operating conditions. Although the technology is proven, the products tend to be costly and troublesome to use. Residual tars can foul gas systems, and the oils produced can be acidic and difficult to store. As a result, few companies have had commercial success with pyrolysis systems (Erickson 1986b).

Liquefaction, the process furthest from commercialization, usually involves taking the products of either gasification or pyrolysis and converting them, through catalytic reactions, to liquid fuels. Medium-BTU gas can be converted in this manner to methanol, a high-octane fuel commonly used in racing cars and as a gasoline additive. No methanol-from-biomass industry has sprung up, however, because methanol is currently made more cheaply from natural gas.

The conversion of wood to gasoline is an interesting avenue of research, because the product is compatible with existing automobiles and existing service stations. In this process, a biocrude oil is first produced through pyrolysis and then upgraded through catalytic reactions to gasoline. Using current technology, gasoline could be produced from wood at an estimated cost of $1.60 per gallon, and the cost is expected to fall to $0.85 per gallon in 20 years (Hubbard 1989a, DOE 1988b). By that time it would almost certainly be less expensive than gasoline derived from petroleum.

Biochemical Conversion: Ethanol and Methane

One of civilization's earliest discoveries was that fermenting corn, wheat, and other grains with yeast produces grain alcohol, or ethanol. In the United States, fuel ethanol became a substantial industry in the early 1980s when high gasoline prices prompted a search for alternative fuel supplies. With the benefit of federal and state tax incentives of $0.60 per gallon and the availability of large grain surpluses as feedstocks, sales of ethanol (mixed with gasoline in a 10 percent blend called gasohol) jumped from 20 million gallons in 1979 to 860 million gallons (equivalent to 0.07 EJ) in 1988 (Finneran 1986b, Dinneen 1989). About 8 percent of gasoline sold in the United States contains ethanol, and in some farm states gasohol has penetrated as much as one-fourth to one-third of the gasoline market (Finneran 1986b). The US ethanol industry, though substantial, is dwarfed by that of Brazil, which produces almost three billion gallons each year from sugarcane (Shea 1988). A major driving force behind the market for ethanol in the United States has been the demand for a clean-burning substitute for lead, an octane-enhancing gasoline additive that is being phased out because of its harmful effects on human health.

Pathways for Biofuels Production

Figure 18. Pathways for the production of biofuels. Adapted from Department of Energy (1988d).

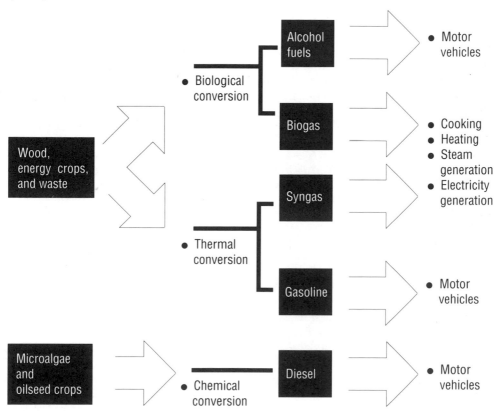

At present, grain surpluses, particularly corn, are the most important feedstocks for ethanol production. However, they will not suffice if the ethanol market grows beyond one to two billion gallons per year. Research is focusing on developing methods of producing ethanol from woody and herbaceous crops (trees, shrubs, and grasses) that could be grown on energy farms. The three main constituents of such plants, cellulose, hemicellulose, and lignin, can now be converted into fuels by various processes including fermentation, hydrolysis, and enzymatic reactions. With over 90 percent of plant components usable, ethanol can be produced from woody biomass at an estimated cost of $1.20-$1.35 per gallon (Hubbard 1989a, Lynd 1989). A price of $0.60-$0.80 per gallon could be achieved by the turn of the century (DOE 1988b, Hubbard 1989a). Because of differences in the volumetric energy content and thermal efficiencies of ethanol and gasoline, ethanol at $0.60 per gallon would be competitive with gasoline at $0.70-$0.80 per gallon (excluding taxes and distribution costs).[2]

One question that is often raised is whether the production of ethanol requires as much or more energy for cultivation, distillation, and other processes than is contained in the fuel, thus diminishing the environmental and economic benefits of ethanol use (Hall *et al.* 1986). While this may be the case when corn is the feedstock, the energy balance appears to be much more favorable for high-productivity methods using lignocellulose plants. It is estimated that the energy input-to-output ratio for ethanol production from such processes is currently about 30 percent, and with foreseeable improvements in technology it could be as low as 15 percent (Lynd 1989, Turhollow 1989).

Anaerobic digestion is another familiar process that has been used for years to treat municipal wastes and sewage, though usually as a means to control air and water pollution, not to extract energy. Anaerobic digestion produces a medium-BTU mixture of methane and carbon dioxide called biogas. The mixture can be purified to yield high-BTU methane, the main component of natural gas.

The majority of anaerobic-digestion plants have been built to collect methane naturally produced in landfills. As of 1988, 50 were in operation with another 44 planned or under construction. Most of the plants produce low-BTU gas, which is often used on-site to generate electricity. In 1988, the LBG plants had a total capacity of 231 MW and generated approximately 0.77 billion kilowatt-hours. High-BTU gas (HBG) is usually sold to natural-gas pipelines. Energy production from both types of plants is estimated to have been about 0.01 EJ in 1988 (Klass 1988).

Anaerobic digestion is also used to convert animal manure, crop residues, sewage, and industrial wastes to energy. About 267 such plants have been built. While their net energy production is uncertain, if run at full capacity they would produce less than 0.005 EJ per year (Klass 1988). Anaerobic digestors have been more successful in less-developed countries. In China, for example, over seven million small digestion systems have been installed (Finneran 1986a).

In general, anaerobic digestion is economical where wastes are available and there is a ready demand for the gas. The costs of purifying the gas (to remove both carbon dioxide and residues from the digestion process) often make it too expensive to be sold to natural-gas pipelines. To make the process more commercial, research has focused on improving the yield and reducing the cost of digestion systems. Scientists are only now beginning to understand the biological processes involved in anaerobic digestion, and experiments are being conducted to develop genetically modified organisms with improved performance. The Department of Energy estimates that the cost to produce pure methane in digestion systems has been reduced from $8/GJ in 1980 to about $5/GJ today, and the cost is expected to drop to $3.50/GJ by the mid-1990s (DOE 1988b). For comparison, the current wellhead cost of natural gas is just under $2 per million BTU but is expected to rise to $3-$4 per million BTU by 1995 (DOE 1988a). Thus, biomass-derived methane could become an economical alternative to natural gas by the mid-1990s.

One promising application of both biogas and syngas is the generation of electricity by gas turbines. Integrated coal-gasification/gas-turbine systems have been intensively studied in the United States as a possible means of generating electricity more cleanly from coal. The systems could be adapted "with modest additional R&D effort" to burn gases derived from biomass (Williams 1989). The capital costs would probably be lower than those for coal because sulfur-removal equipment would not be necessary. The California Energy Commission projects that utility-owned biomass-gasification/gas-turbine systems completed in 1992 could produce electricity at a highly competitive cost of 6-11¢/kWh (CEC 1988a).

Plant-Oil Fuels

Another method of producing liquid fuels, mainly diesel fuel, is to extract and refine oils from plants. Farmers in the United States have at times burned sunflower oil, soybean oil, and other plant oils in their diesel tractors, though fouling has been a problem (Layton 1989). A more promising crop for producing large quantities of fuel oil, because it can be grown in off-seasons, is winter rapeseed. The present cost of rapeseed oil is estimated to be about $2.75 per gallon, and this could fall to $1.50 per gallon, including distribution costs, by 1995 (DOE 1988b). The production of fuel oil from microalgae is at a much earlier stage of development and will probably not become commercially available for several decades (Turhollow 1989).

Environmental Impacts

All energy technologies have some impact on the environment, and biomass technologies are no exception. The environmental impacts of widespread biomass use are among the most important uncertainties concerning this resource. Unfortunately, the impacts have not, in general, been the subject of careful or systematic analysis. This report cannot substitute for such analysis, which must be conducted by experts in a number of fields. However, some of the important issues and their implications are discussed briefly below.

Global Warming

The widespread substitution of biomass for fossil fuels would substantially reduce greenhouse-gas emissions. Although carbon dioxide is emitted when biomass is burned, it is removed from the atmosphere in almost equal quantities if the biomass is continually regenerated by new growth. Carbon dioxide is also emitted by fossil fuels used in the production of biomass energy (e.g., for transportation and fertilizers), but in successful biomass technologies, the fossil-energy input should be much less than the renewable-energy output, so that the overall contribution to global warming would still be small. If the energy input came from renewable sources, there would be no net contribution to global warming.

Air Pollution

A more difficult issue is the effect of biomass combustion on air pollution. Although biomass is generally low in sulfur typically found in coal, its direct combustion can produce other pollutants, such as particulates (fine particles of soot and ash). The need to control pollutant emissions will inevitably add to the costs of some biomass technologies. The problem may be particularly severe for facilities burning raw municipal wastes, which can contain toxic metals and other harmful compounds. Mass-burn facilities also produce toxic ash that must be safely disposed of. For these reasons, it may be environmentally preferable to recycle most municipal wastes rather than burn them.

Burning methanol and ethanol instead of gasoline would substantially reduce emissions of most automobile pollutants, particularly unburned hydrocarbons and other volatile organic compounds (VOCs) that react in the presence of heat and sunlight to produce ozone. In advanced cars designed specifically for pure-alcohol fuels, it has been shown that VOC emissions would be reduced 85-95 percent and carbon-monoxide emissions 30-90 percent (EPA data cited by Lynd, 1989). In unmodified cars the reductions would not be as large, and in either case emissions of nitrogen oxides (a source of acid rain) would not be greatly affected. The only pollutants that would be significantly increased by burning ethanol and methanol are aldehydes. Opinions differ over what level of aldehyde emissions would be dangerous and whether the emissions could be effectively controlled with catalytic converters. Of course, burning biomass-derived gasoline instead of conventional gasoline would provide no air-pollution benefits.

Implications for Agriculture and Forestry

It is a perhaps surprising fact that growing trees and plants for energy could benefit cropland and farmers in the United States. Energy crops could provide a steady, supplemental income for farmers in off-seasons or on unused land without requiring much additional equipment and without competing against food crops. Moreover, woody and herbaceous energy crops could help stabilize marginal or overused cropland afflicted by excessive soil erosion and chronic flooding. Trees would grow for several years before being harvested, and their roots and leaf litter would help hold the soil; coppicing (self-regenerating) varieties would minimize the need for disruptive tilling. Soil loss from perennial energy crops such as switchgrass would also be much lower than that from annual food crops (Wright et al. 1989).

But if not properly managed, energy farming could have harmful effects as well. While most energy crops would be grown using far less herbicides and pesticides than conventional food crops, large-scale energy farming—at least as it is now envisioned—would nevertheless entail a substantial increase in chemical use because more land would be under cultivation. It would be preferable, even though crop yields might be reduced, to grow "organic" energy crops using few or no toxic chemicals. Care must also be taken not to remove too many soil nutrients when agricultural and forestry residues are collected and burned. (On the other hand, some residues and ash left over after biomass combustion or fuel conversion could be returned to the soil as a fertilizer.)

Increasing the amount of wood energy extracted from forests could have positive or negative effects. It could provide an incentive for the forest-products industries to manage their resources more efficiently, to the benefit of forest health. But it could also provide an excuse to exploit new forest areas in an unsustainable fashion. Unfortunately, commercial forests have not always been soundly managed, and many people will view with alarm the prospect of even greater forest exploitation. Their concerns must be met by tighter federal and state controls on forestry practices following the principles of "excellent" forestry advocated by Robinson (1988) and others. If such principles are followed, it should be possible to extract energy from forests indefinitely, without damaging their scenic or natural value.

Conclusions and Near-Term Prospects

Biomass now accounts for about 4-5 percent of US energy supply, and a number of technologies are being developed that will soon open new opportunities for the exploitation of this resource. The most promising in the near term are processes that convert wood and other plant matter to ethanol and agricultural and other wastes to gaseous fuels. With continued research and development, both types of fuels could become competitive with fossil fuels in the 1990s.

Biomass could eventually reduce US fossil-fuel consumption by as much as a third and, if converted to liquid fuels, replace some 30-90 percent of the gasoline and diesel fuel currently consumed for transportation. Such widespread use of biomass would substantially reduce emissions of greenhouse gases and pollutants. However, the exploitation of the biomass resource must be carefully managed to avoid other harm to the environment. While it appears that in most cases the risks can be minimized through conventional pollution-control strategies and intelligent resource management, some environmental issues—particularly those associated with the large-scale cultivation of energy crops—need to be studied more carefully.

In the present climate of low fossil-fuel prices and minimal government support, however, biomass is not likely to make a much greater contribution to US energy supply until after the turn of the century. Funding for biofuels research and development has dropped from $49.7 million in fiscal 1981 to $13.4 million in fiscal 1989 (Sissine 1989). Klass (1988) projects that, with no change in government policy, biomass-energy consumption could grow to 4.37 EJ in 2000, just 40 percent more than consumption in 1987 and not enough to significantly affect fossil-fuel use (see Figure 17).

With additional government support and market incentives, however, the biomass contribution could grow much more quickly. Projections of biomass consumption in 2000 under "enhanced" scenarios range from 8.4 EJ (Rader 1989) to 15.3 EJ (Klass 1988). In our opinion, a goal of 6-8 EJ, or 7.5-10 percent of current primary-energy consumption, is readily achievable. This would require, for example, the conversion of one-fourth to one-half of available forestry and agricultural wastes to energy and the cultivation of energy crops on 20-40 million acres of presently idle cropland.

Notes

1. Much higher biomass yields can be expected in some areas, but the estimate used here (5 dry tons/acre/year) represents an average appropriate for large land areas using current cultivation techniques and plant strains. The assumed efficiency of conversion to ethanol—50 percent—reflects the state of the art, though efficiencies as high as 65 percent are theoretically possible.

2. Ethanol has a heating value of about 80 megajoules (MJ) per gallon, whereas gasoline has a heating value of about 123 MJ per gallon. However, the thermal efficiency of ethanol-powered vehicles is approximately 10-20 percent higher than that of gasoline-powered vehicles (Lynd 1989), implying that one gallon of ethanol will substitute for 0.77-0.84 gallons of gasoline.

Hydroelectric Power

Rivers played an integral role in the early development of the United States, and thus it is perhaps natural that hydroelectric power became one of the most important sources of electricity in the early part of this century, after the first plant was completed at Niagara Falls in 1878. Even today, rivers provide 10-12 percent of US electricity supply, depending on demand and rainfall, or as much is produced by 40-50 nuclear power plants.

There remains substantial room for further hydroelectric development in the United States. It is uncertain, however, how much of the remaining resource will actually be developed. The era of large dam building, which saw the construction of such behemoths as the Hoover Dam (1,455 MW) and the Grand Coulee Dam (6,180 MW), is drawing to a close, while small-scale hydroelectric development has become increasingly controversial because of concerns for the environment. Although hydropower will continue to be an important electricity source for many decades, significant growth of hydroelectric capacity, beyond the expansion and upgrading of existing facilities, seems unlikely.

The Hydroelectric Resource

The United States is blessed with many rivers and streams capable of supporting hydroelectric power plants. The Federal Energy Regulatory Commission (FERC) has cataloged 7,319 sites in the United States which, if developed, could support 146,900 MW of hydroelectric capacity. Of this potential, 2,010 sites of 70,800 MW total capacity have been developed, and another 174 projects of 1,400 MW capacity are under construction, 20 of which represent expansions of existing facilities. Thus, in theory, the potential exists for the United States to just about double its present hydroelectric capacity (FERC 1988). Over the past five years, hydroelectric power plants have produced an average of 273.2 billion kilowatt-hours annually, displacing 3.24 exajoules (EJ) of primary energy (EIA 1989b); if all available sites were developed, annual electricity production could reach an estimated 534.9 billion kilowatt-hours and displace 5.8 EJ of primary energy (FERC 1988).

In fact, the resource that could actually be exploited is probably much smaller. Environmental laws, in particular, place severe restrictions on new hydroelectric development. The 1968 National Wild and Scenic Rivers Act and other federal legislation preclude the building of facilities on stretches of virgin rivers representing an estimated 40 percent of the nation's remaining (undeveloped) hydroelectric potential, while an additional 18 percent are under a development moratorium until their final status is decided (FERC 1988). Assuming that half of the latter sites are eventually placed under protection, then no more than half of the remaining potential, or about 37,000 MW, will be available for development.

Many of the remaining opportunities for hydroelectric development involve upgrades or expansions of existing facilities. The capacity of the Grand Coulee facility, for example, was recently increased by over 2,000 MW with the addition of three new turbines and the rewinding of existing turbines. In addition, there are over 3,300 small hydroelectric facilities that were abandoned in the 1950s and 1960s, some of which might be returned to operation, as well as thousands of small, nonhydroelectric dams that could potentially be converted to electrical generation (FERC 1988, Rogers 1989).

With the development of domestic rivers constrained, the United States is likely to turn to hydroelectric power imported from Canada. Some observers expect imports to almost double over the next 10 years (Rader 1989). Canadian hydropower offers no long-term solution for US electricity needs, however: the Canadian resource, though large, is just over half the US resource, and it seems likely that environmental and economic concerns will prevent the development of much of it.

Pumped hydroelectric storage is often counted as hydroelectric power, although it actually only stores electricity (by pumping water from a lower reservoir to a higher reservoir) for generation in peak periods. As of 1988, about 17,100 MW of pumped-storage capacity were in operation at 37 sites, with another 2,000 MW at three sites under construction (FERC 1988). FERC has identified 38 sites that could be developed with a potential capacity of 17,100 MW, and many more potential sites exist but have not yet been surveyed. Pumped hydroelectric storage could be extremely important to the future development of solar and wind power, as it is probably the only large-scale electrical-storage technology that is presently cost effective and widely available (OTA 1985).

Technologies and Economics

Hydroelectric power plants, both large and small, are a proven technology. In large plants, a reservoir is created by the damming of a river, and the water in the reservoir is allowed to fall up to hundreds of feet through a turbine, generating electricity. (As a rough rule of thumb, one gallon of water per second falling one hundred feet can generate one kilowatt of electrical power.) Small-plant designs vary considerably. Some are dams with a head (the distance between the top of the reservoir and the bottom of the dam) of less than 60 feet. Dams are not always necessary, however; turbines can be placed in pipes in midstream, with the pipes diverting all or only part of the stream flow. One advantage of such axial-flow designs is that they can reduce the impacts on fish and aquatic life downstream of the plant (discussed below).

Traditionally, hydroelectric power has been one of the least expensive sources of electricity. Although its cost advantage is no longer so great, it remains on the whole an economical technology. The cost of hydroelectric power varies widely depending on such factors as the size of the plant, the design, the capacity factor (typically 40-50 percent, as hydroelectric plants are most often used to provide intermediate and peak, not baseload, electricity), the costs of mitigating environmental impacts, and the availability of transmission lines. As is the case with most renewable energy sources, initial capital costs are relatively high, whereas operations and maintenance costs are low. Large plants typically cost between $500 and $2,500 per installed kilowatt capacity, while the cost of small plants ranges from less than $1,000 per kilowatt to more than $6,000 per installed kilowatt and average around $2,000 per kilowatt (Shea 1988). Operations and maintenance costs at privately owned plants average about 0.2¢/kWh, while at larger government-owned plants they average about half as much (EIA 1988b).

Official estimates of the cost of electricity from recently installed hydroelectric capacity are 3-6¢/kWh (Rader 1989). These figures include relatively inexpensive retrofits and upgrades of existing facilities, however. The California Energy Commission estimates that the cost of electricity generated by new small hydroelectric plants completed in 1992 will be 8-20¢/kWh (CEC 1988a, converted to 1989 dollars). Even then, hydroelectric power should continue to be competitive with most other technologies for generating intermediate and peak power.

Besides low cost, hydroelectric power offers the advantage that it can easily be switched on and off and so is ideal for meeting peak and emergency demand. Hydroelectric power generation is affected by the seasons and precipitation, however. (The drought of 1988 contributed to a 25 percent drop in national hydroelectric output.) Water resources may become more scarce in the future if predictions of increased summer dryness caused by global warming prove correct.

Despite the apparent economic and practical advantages of hydroelectric power, its development has become increasingly problematic. Large-scale hydroelectric development is being phased out—except for the expansion and upgrading of existing facilities—because most remaining undeveloped sites suitable for large dams are under federal protection. To some extent, the slack has been taken up by a revival of small-scale development sparked by PURPA and federal tax credits. But small-scale hydro development has not met early expectations. Between 1984 and 1988, just 650 MW of small-scale hydro capacity (defined as less than 30 MW each) were added, and as of 1988, small hydro plants accounted for 7,235 MW, or about 10 percent, of US hydroelectric capacity (FERC 1988).

Environmental Impacts

While declining fossil-fuel prices and the elimination of renewable energy tax credits in 1985 have both acted to slow hydroelectric development, public opposition and environmental regulation have played an important part. The small-scale hydroelectric industry believes current regulations are an excessive burden, one that adds 30 percent to the average cost of new small facilities (Rogers 1989). The industry's problems are likely to become worse; the 1986 Electric Consumers Protection Act, which was designed to place consideration of fish and wildlife interests on an equal footing with power generation, has, in the words of one industry representative, "ensured that new small hydro will not be built" (Rogers 1989).

That environmental regulations would constrain hydroelectric development in the United States is ironic, since hydroelectric plants produce no air pollution or greenhouse gases. Yet hydroelectric projects can affect the environment in other important ways. The impacts of very large dams are so great that there is little chance that any more will be built in the United States. (Large hydroelectric projects are being pursued in many less-developed countries, however, often with inadequate attention to their human and ecological impacts.) Depending on the location, reservoirs created by large dams inundate forests, farmland, wildlife habitats, and scenic areas. In addition, dams cause radical changes in river ecosystems. Sediments bearing nutrients accumulate in reservoirs instead of being carried downstream, and changes in river-flow rate, temperature, and oxygen content alter the balance of plant and fish life. Reservoir evaporation increases salt and mineral content, a serious problem for the Colorado and other mineral-rich rivers. Dams also block fish migration and destroy spawning grounds, resulting in losses for commercial fisheries and sport fishing (Shea 1988, Kendall and Nadis 1980).

All of these problems can be of concern with small hydroelectric dams and reservoirs, as well. Some of the impacts can be mitigated by such methods as building screens to divert fish from deadly turbine blades and "ladders" to allow fish to migrate over dams, and maintaining minimum river-flow rates, although the effectiveness of these steps has not been firmly established (Shea 1988). Axial-flow or run-of-the-river systems, which create no reservoir, generally have the fewest impacts, although the maintenance of adequate stream-flow rates and other problems can still be of concern.

Conclusions and Near-Term Prospects

Hydroelectric power makes an important contribution to the nation's electricity supply and will continue to do so for many decades. However, the opportunities for substantial further growth of hydroelectric capacity appear increasingly limited. Almost half of the nation's hydroelectric resource has been developed, and development of much of the rest faces major hurdles. While US hydroelectric capacity will continue to grow in coming years, it appears unlikely to surpass 100,000 MW—40 percent more than existing capacity—or displace more than the primary-energy equivalent of about 4 EJ.

The most important policy question that should be addressed concerns whether regulations governing small hydroelectric development should be streamlined or loosened. Streamlining, or simplifying, the licensing process could be beneficial—as long as public input is not restricted—if it helps deserving projects go rapidly forward while preventing undeserving ones from hanging on indefinitely. But easing environmental standards governing new development raises more serious questions. The hydroelectric-power industry claims that it suffers under an excessive regulatory burden, while fossil-fuel industries—whose environmental impacts, including global warming, are arguably more severe—are given virtually a free ride. Though correct as far as it goes, this view fails to recognize the extraordinary value society places on the preservation of diminishing natural resources and wildlife. There is a clear consensus that a good portion of the nation's river system should be kept as much as possible in its natural state for the enjoyment and appreciation of present and future generations. Despite growing concern about global warming and pollution, it is unlikely that this consensus will change in the future. Nor is it easy to argue, given the relatively small contribution hydropower could make to reducing fossil-fuel use, that it should.

Chapter seven

The Path to a Renewable Future

As the foregoing chapters have shown, wind, solar, and biomass technologies have made remarkable technical progress in the past 10 years, and several are now or soon could be competitive with fossil fuels. Yet ironically, their progress toward commercial acceptance is slower now than at any time since the 1970s. Most renewable-energy industries made their greatest market gains in the early 1980s. Since then, their growth has slowed considerably and in many cases has reversed.

Things could get worse before they get better. A survey of 108 renewable energy companies conducted by the Investor Responsibility Research Center found that, although 1,280 MW of renewable electrical-generation capacity are expected to be added in 1989, only 211 MW are planned or under construction for 1993 (Williams and Porter 1989). Many companies have nearly exhausted the long-term, favorably priced contracts awarded by some utilities in the early 1980s. Although the market picture is likely to improve toward the end of the 1990s, a Department of Energy forecast nevertheless suggests that renewable energy sources will account for about 12 percent of projected US energy supply in 2010, compared to 7.6 percent today—hardly an impressive leap, and not enough to affect fossil-fuel use and global warming in a significant way (DOE 1988a).

Market Barriers

Understanding the causes of the slowdown in renewable-energy industries will be crucial to developing policies that will encourage their growth in the future. Chief among the causes have been an economic climate favoring the energy status quo, a sharp decline in government support, and a market that continues to be insensitive to the environmental and social benefits of renewable energy sources.

The Economic Climate

The economic climate for renewable-energy industries has changed dramatically since 1985. The world is now experiencing a glut of low-cost fossil fuels, especially oil, brought about by greater-than-expected energy conservation by industrialized nations and the discovery of new oil and gas fields in the 1970s and 1980s. As a result, the price of crude oil is lower in real terms today than it has been at any time since 1973 and, at $15-$20 per barrel, is about half what it was at its peak in 1981. Natural-gas prices, too, have fallen almost 50 percent since their peak, and coal prices—normally immune to the sharp price fluctuations of petroleum and gas—have joined the downward trend. Lower fuel costs have also meant lower prices for electricity (EIA 1989a).

In this climate, there is little incentive for the private sector to seek or develop alternatives to fossil fuels. Although in the long run this is sure to change, it may take several years for excess fossil-fuel supplies to be absorbed by rising demand, and thus prices are likely to remain relatively low for some time. The Department of Energy forecasts that the real price of crude oil will not return to its 1981 level until after the turn of the century (DOE 1988a). (See Figure 19.)

Ironically, increased energy efficiency in the United States and other countries would tend to keep fossil-fuel prices low, thus making it more difficult for renewable energy technologies to find a market. On the other hand, oil and gas prices could rise much more quickly than current projections indicate if, for example, a war forced the shutdown of major oil fields or supply routes in the Middle East. This is a major risk of the United States' increasing reliance on imported oil (Yergin 1988).

The growth of renewable-energy industries is further slowed by other, more subtle market barriers. The lack of familiarity of most people with renewable-energy technologies has discouraged interest in them and made it more difficult for companies to obtain financing for their projects. The poor economic performance of some early projects (most notoriously wind farms) has also given them something of a bad name in financial circles. Ironically, even the relatively small size of some renewable-energy projects has at times worked against them, since major lenders normally do not deal with small loans.

Our society's emphasis on short-term profits also works to the disadvantage of renewable-energy technologies, many of which cost more up front than fossil-fuel technologies but save money in the long run because of lower fuel costs. Although the use of 30-year levelized costs adopted in this report should correct for this bias, most energy decisions, even at the corporate level, are not made on this basis. Private consumers often demand a payback period of just two to three years for their investments. Moreover, even levelized costs often do not give sufficient weight to the risk of rapid, unexpected increases in fossil-fuel prices like those that occurred in 1973 and 1978–1979.

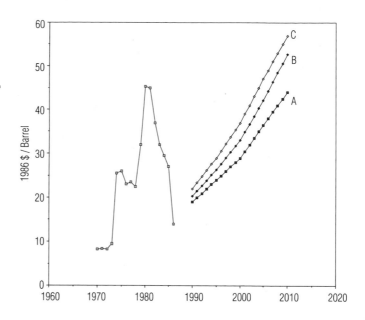

Figure 19. Historical and projected future oil prices, 1970-2010. Curves represent (A) low economic growth, (B) the reference case, and (C) high economic growth. Source: Department of Energy (1988a).

Federal Support

Another factor slowing the growth of renewable-energy industries has been the decline of federal support, especially the elimination of renewable-energy tax credits. From 1978 to 1985, combined federal investment and solar-energy tax credits amounted to 25 percent, and in California and other states they were supplemented by state tax credits as well. These credits helped attract risk capital to renewable-energy industries and correct for the inherent biases of energy markets. Today only a reduced 10 percent federal tax credit remains, subject to annual approval by the Congress, for solar, ocean-thermal, and geothermal business investments. Tax credits for wind, biomass, and residential energy investments—the last a key prop supporting the residential solar-collector industry—were eliminated in 1985.

Federal funding of renewable-energy research and development has declined as well, from a peak of $718.5 million in fiscal 1980 to $114.5 million in fiscal 1989 (including geothermal and ocean-energy research), resulting in much slower advancement of technologies (see Figure 20). Government/industry demonstration projects, which in the past have provided valuable experience with, and markets for, new technologies, are for the most part no longer being pursued or have been greatly scaled back.

In part because of declining federal support, the United States has seen its edge in renewable-energy technologies eroded by foreign competitors (Paul 1989, Victor 1989b). Although the United States continues to spend more than any other country on renewable-energy research and development, its per-capita spending in 1986 was only one-third that of Sweden (Shea 1988). The once-dominant US photovoltaic industry has given up almost two-thirds of the world photovoltaic market to Japanese and European manufacturers, and several countries, such as Greece, Sweden, and Israel, are aggressively developing and commercializing solar technologies for homes and buildings. Ominously, the United States has become a net importer of solar-thermal and wind systems (Sissine 1989).

The Hidden Costs of Energy

One of the principal barriers to the commercialization of renewable-energy technologies is that current energy markets ignore, for the most part, the social and environmental costs and risks associated with fossil-fuel use. In effect, relatively harmful energy sources, like coal, are given an unfair market advantage over relatively benign sources, like wind power. Some conventional energy sources are also heavily subsidized, directly or indirectly, by the government. If these external, or hidden, costs were included in the price of energy, renewable-energy technologies would be in a far better position to compete with fossil fuels.

Social and environmental costs take many forms. Coal-fired power plants and automobiles, for example, damage human health and reduce agricultural production by contributing to acid rain and air pollution. Overreliance on imported oil makes the United States vulnerable to supply disruptions and price increases and necessitates large naval fleets to defend Persian Gulf shipping. Nuclear power creates a public-safety hazard as well as a radioactive-waste disposal problem.

While it is extremely difficult to quantify these costs—and some cannot be quantified—nevertheless they appear to be substantial. One study estimated that coal and nuclear plants both carry external costs (including, in the case of coal, the effects of global warming) of at least 2-3¢/kWh, to which, the authors argue, a premium must be added for costs not yet quantified (Cavanagh et al. 1982). Similarly, a West German study concluded that the external costs (not including global warming) of electricity from fossil-fuel plants are in the range of 2.4-5.5¢/kWh, while those of nuclear plants are 6.1-13.1¢/kWh (Hohmeyer 1988).

Government subsidies in support of energy technologies are another hidden cost that is often ignored. Subsidies include tax breaks as well as direct expenditures, such as for research and development, cleanup of environmental accidents, regulatory agencies, and other programs. One study estimated that federal energy subsidies amounted to $44 billion in 1984 (Heede et al. 1985). The pattern of subsidies indicated a massive investment in support of traditional energy technologies, rather than an investment in developing new technologies for the future. Over 90 percent of 1984 subsidies went to established energy sources—principally coal, oil, natural gas, and nuclear power. Emerging renewable energy technologies, excluding hydroelectric power, received less than 4 percent of the total.

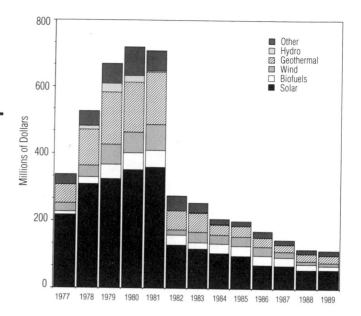

Figure 20. Department of Energy funding for renewable-energy research and development, fiscal 1977-1989. "Other" includes funding for ocean-energy research, technology transfer, resource assessment, and other programs. Source: Sissine (1989).

PURPA and Utility Regulation

The 1978 Public Utility Regulatory Policies Act, or PURPA, has provided a powerful impetus for the development of biomass, wind, small-hydroelectric, and other renewable sources of electricity. PURPA forces utilities to purchase electricity from qualifying facilities (QF) at a price equal to the "avoided cost" of utility generation. (The avoided cost is the price a utility would have to pay for electricity if it did not buy it from an independent producer.) A qualifying facility can either be a cogenerator or a small power facility. A cogenerator makes use of the waste heat from a generating plant to raise the overall efficiency of energy production. A small power facility is defined as one whose capacity is no greater than 80 MW and whose primary energy source is renewable.

PURPA was intended to help diversify the nation's energy sources, promote renewable energy, and reduce costs for consumers, and to a considerable degree it has succeeded. However, since 1985, because of the changing economic climate, the elimination of renewable-energy tax credits, and increased competitive bidding on contracts, PURPA has strongly favored cogenerators (most of them burning natural gas) over renewable energy sources. Only 12 percent of proposed new independent power plants in 1987 were renewable-based, down from 29 percent the previous year (Shea 1988).

Because of their higher efficiency and lower pollutant emissions, natural-gas cogenerators are much preferable to conventional coal-fired power plants. They should not, however, push renewable energy sources aside. US reserves of natural gas are destined to dwindle in coming years, so that imports and prices can be expected to rise. In addition, natural-gas combustion contributes to global warming, albeit to a lesser degree than coal and oil combustion.

In March 1988, the Federal Energy Regulatory Commission (FERC) proposed changes in PURPA regulations that would make life even more difficult for small power producers. The changes, if approved, would essentially eliminate the favored status of qualifying facilities and put them on an equal competitive footing with other independent power producers and utility-owned facilities. By making economics once again the chief factor in utility decisions, the proposed new rules are heavily biased toward conventional generating technologies, which are for the most part far more damaging to the environment than renewable sources (Rader 1989).

Policy Recommendations

If the current market gives insufficient weight to the environmental and social benefits of renewable energy sources, then federal, state, and local governments must step in. Governments can have a decisive influence on energy choices, and the budget burden need not be large. For renewable-energy technologies, many of which are on the edge of commercialization, government actions can be especially cost effective. The key is to find the policy levers that have the greatest influence on the development of renewable energy sources, and pull them.

We urge, as an initial step, the adoption of the following six federal policies. We believe that the implementation of these policies could lead to a doubling of the energy production from renewable energy sources by 2000 and a corresponding decrease of 5-10 percent in fossil-fuel consumption and carbon-dioxide emissions.

1. The federal government should reinstitute renewable-energy tax credits.

Tax credits are a very effective means of boosting investment in emerging technologies, although their image has suffered somewhat because of the poor performance of some early renewable-energy projects. The problems, due in part to developers seeking quick profits on virtually tax-free investments, can be largely avoided in the future by tying the credits to performance and energy output, rather than to installed capacity; alternatively, or perhaps in addition, the federal government could certify renewable-energy systems to ensure that they meet strict standards of reliability and performance. For residential or small-business investments, an up-front tax rebate that reduced the effective purchase price of solar collectors and other small-scale technologies would be most effective.

Tax credits would probably result in a reduction of several billion dollars per year in government revenue by the year 2000.[1] From an economic standpoint, this loss would be justified if it reflected the reduction in social and environmental costs due to the displacement of fossil fuels by more environmentally benign energy sources. As a practical matter, however, a corresponding source of revenue might be necessary to offset the credits and relieve any burden on the federal budget. Additional revenue could come in the form of taxes on carbon-dioxide emissions or fossil fuels, as discussed below, or a corresponding reduction in subsidies for fossil fuels and nuclear power.

2. Modest taxes should be levied on fossil-fuel consumption, with part of the revenue used to fund research and development of renewable energy sources and to offset the costs of renewable-energy tax credits.

Increased taxes on gasoline and other fossil fuels are often thought of as a tool for discouraging consumption and encouraging efficiency. However, in order to reduce consumption appreciably, taxes would probably have to be set so high as to be politically impractical, at least in the near future. In Europe, for example, taxes more than double the price of gasoline to $2.50-$3.00 per gallon, yet the overall fuel economy of cars in Europe is only three miles per gallon higher than that in the United States (EA 1989).

On the other hand, modest taxes could help pay for the development of alternative energy technologies. A 10¢/gallon federal gasoline tax on top of existing federal and state taxes would raise about $10 billion in revenue, and if the tax were distributed among all fossil fuels it would be correspondingly lower. Since taxes would have a disproportionate effect on the poor, there would have to be adjustments to the tax code for low-income taxpayers.

3. PURPA should be modified to require electric utilities to take into account the social and environmental costs of energy technologies when contracting for new capacity.

This proposal is directed at electric utilities for two reasons. First, they are probably the most important near-term market for renewable-energy technologies. Electricity demand is growing at a faster rate than overall energy consumption, and it is likely that a large amount of additional electrical-generation capacity—as much as 100,000 MW, according to Energy Information Administration forecasts (EIA 1989c)—will be required by 2000. In addition, electric utilities are the only major economic sector under tight federal and state regulation and thus are the most amenable to government influence.

Only a few states, including Wisconsin, California, and New York, are seriously considering or have instituted regulations requiring electric utilities to consider the environmental and social costs of energy, although more states are likely to follow (Burkhart 1989, Cohen 1989). Modifying PURPA would have the effect of requiring all states to adopt such regulations. The social and environmental costs would be assessed on new, not existing, capacity. But to avoid giving electric utilities added incentive to extend the lives of existing plants—many of which burn coal—the regulations should also be applied to life-extension plans.

One objection that is sure to be raised is that because it is so difficult to calculate the environmental and social costs of energy, almost any estimates proposed by electric utilities or others will be open to challenge. Given the uncertainties, in fact, there will be a natural tendency for the external costs to be set too low, at least at first. Nevertheless, nearly any scheme would be preferable to the present one, in which almost no consideration at all is given to the environmental and social damage caused by fossil fuels.

4. Research-and-development funding for renewable energy should be steadily increased by 20-30 percent per year for the next several years, approaching $1 billion by 2000.

Greater emphasis should also be placed on near-term commercialization, including cost-sharing efforts, utility demonstration projects, and market-oriented research on product durability, efficiency, and reliability to enhance quality and consumer acceptance.

5. The government should purchase renewable-energy technologies for its own facilities, where possible, as a way to foster greater production and economies of scale.

For example, the government could purchase photovoltaic cells for remote sites, as the Coast Guard has already done for navigation buoys; government agencies with large car and truck fleets could establish biomass-fuel demonstration programs; passive- and active-solar technologies could be routinely incorporated in new government buildings.

6. The government should expand efforts to encourage exports of renewable-energy technologies.

The international market for renewable-energy technologies is potentially enormous. Less-developed countries, in particular, are eager to expand their citizens' access to basic necessities and modern conveniences. Unfortunately, the Committee on Renewable Energy, Commerce and Trade (CORECT), set up by the federal government in 1981 to coordinate the promotion of renewable energy technologies in export markets, has been cut drastically in the past several years, and today its annual budget is a token $1 million. This program should be expanded significantly. In addition, the Agency for International Development (AID) can play an important role through its development grants and loan programs, and the United States can use its influence with the multilateral development banks, like the World Bank, to encourage environmentally beneficial programs based on renewable energy.

Notes

1. A rough estimate of the revenue loss can be made by assuming that the tax credits subsidize renewable-energy technologies at an average rate of $1 per gigajoule. An additional 8 exajoules of renewable energy would consequently cost the government $8 billion per year.

References

American Wind Energy Association (AWEA). 1989a. *Wind Energy Has Come of Age in California.* Washington, DC: AWEA.

———— 1989b. "Response of American Wind Energy Association to 'Wind Energy Development in California.'" Washington, DC: AWEA.

———— 1988a. "Wind Energy Costs Drop." Collection of viewgraphs. Washington, DC: AWEA.

———— 1988b. *Windletter.* Issue #7.

Bain, Don. 1988. "Pacific Northwest Wind Energy Planning." Oregon Department of Energy.

Bath, Thomas D. 1989. Manager, Analysis and Evaluation Office, Solar Energy Research Institute. Testimony before the Senate Committee on Energy and Natural Resources. March 14.

Bleviss, Deborah Lynn. 1988. *The New Oil Crisis and Fuel Economy Technologies.* New York: Quorum Books.

Boes, Eldon. 1989. Sandia National Laboratory Photovoltaic Technology Division. Personal communication.

Bolin, Bert, Bo R. Doos, Jill Jager, and Richard A. Warrick. 1986. *The Greenhouse Effect, Climatic Change, And Ecosystems.* New York: J. Wiley and Sons.

Broad, William J. 1988. "Experts Call Reactor Design 'Immune' to Disaster." *The New York Times.* November 15.

Burkhart, Lori A. 1989. "External Social Costs as a Factor in Least-Cost Planning— An Emerging Concept." *Public Utilities Fortnightly.* August 31.

California Energy Commission (CEC). 1988a. *Energy Technology Status Report.* Sacramento: California Energy Commission.

———— 1988b. *Results from the Wind Project Performance Reporting System: 1987 Annual Report.* Sacramento: CEC. August.

———— 1987. *Relative Cost of Electricity Production.* P300-86-006. Sacramento: CEC.

———— 1985. *Commercial Status of Electrical Generation Technologies.* P300-85-003. Sacramento: CEC.

Cavanagh, Ralph, Margie Gardner, and David Goldstein. 1982. "A Model Electric Power and Conservation Plan for the Pacific Northwest." (Appendix 2.) San Francisco: Natural Resources Defense Council.

Chandler, William U., Howard S. Geller, and Marc R. Ledbetter. 1988. *Energy Efficiency: A New Agenda*. Washington, DC: American Council for an Energy-Efficient Economy.

Cohen, Armond. 1989. Senior Attorney, Conservation Law Foundation. Personal communication.

Cook, James. 1989. "Warming Trend." *Forbes*. February 20, p. 68.

Davenport, Roger L. 1989. Science Applications International Corporation. Personal communication.

Department of Energy. 1989a. *Solar Buildings Program Summary: Fiscal Year 1988*. DOE/CH10093-47. Washington, DC: Government Printing Office.

————— 1989b. *Solar Thermal Program Summary: Fiscal Year 1988*. DOE/CH10093-45. Washington, DC: GPO.

————— 1989c. *Photovoltaic Energy Program Summary: Fiscal Year 1988*. DOE/CH10093-40. Washington, DC: GPO.

————— 1989d. *Energy Storage and Distribution Program*. DOE/CH10093-51. Washington, DC: GPO.

————— 1989e. *Wind Energy Program Summary: Fiscal Year 1988*. DOE/CH10093-41. Washington, DC: GPO.

————— 1988a. *Long Range Energy Projections to 2010*. DOE/PE-0082. Washington, DC: GPO.

————— 1988b. *Programs in Renewable Energy: Fiscal Year 1989*. DOE/CH10093-38. Washington, DC: GPO.

————— 1988c. *Investing In Success*. Photovoltaic Energy Technology Division. Collection of viewgraphs. Washington, DC: GPO. November.

————— 1988d. *Five-Year Research Plan 1988-1992: Biofuels and Municipal Waste Technology Program*. DOE/CH10093-25. Washington, DC: GPO.

————— 1988e. *Five-Year Research Plan 1989-1993: Solar Building Technologies Program*. Washington, DC: GPO.

———— 1987. *Five-Year Research Plan 1987-1991: National Photovoltaics Program.* DOE/CH10093-7. Washington, DC: GPO.

Department of Transportation (DOT). 1989. Office of Public Information. Personal communation.

Dinneen, Robert. 1989. Renewable Fuels Association. Personal communication.

Dodge, D.M., and R.W. Thresher. 1989. "Wind Energy." *Assessment of Solar Energy Technologies.* Boulder, CO: American Solar Energy Society. May.

Electrical World. 1989. "The Great, Light-Water Hope." *Electrical World.* October.

Energy Information Administration (EIA). 1989a. *Monthly Energy Review.* DOE/EIA-0035(89/06). Washington, DC: GPO. June.

———— 1989b. *Annual Energy Review 1988.* DOE/EIA-0384(88). Washington, DC: GPO.

———— 1989c. *Annual Energy Outlook 1989.* DOE/EIA-0383(89). Washington, DC: GPO.

———— 1988a. *Commercial Nuclear Power 1988: Prospects for the United States and the World.* DOE/EIA-0438(88). Washington, DC: GPO.

———— 1988b. *Historical Plant Cost and Annual Production Expenses for Selected Electric Plants 1986.* DOE/EIA-0455(86). Washington, DC: GPO.

Environmental Action (EA). 1989. "RE: Sources." *Environmental Action.* July/August.

Environmental Defense Fund (EDF). 1985. *To Burn Or Not To Burn: The Economic Advantages of Recycling Over Garbage Incineration for New York City.* New York: EDF.

Environmental Protection Agency (EPA). 1989. *Policy Options for Stabilizing Global Climate.* Report to Congress (Draft). February.

———— 1988. *The Potential Effects of Global Climate Change on the United States.* Report to Congress (Draft). October.

Erickson, Jeffry J. 1986a. "Bioenergy: Direct Combustion." *Energy Innovation: Development and Status of Renewable Energy Technologies 1985.* Washington, DC: Solar Energy Industries Association.

———— 1986b. "Bioenergy: Gasification, Pyrolysis, and Liquefaction." *Energy Innovation: Development and Status of Renewable Energy Technologies 1985.* Washington, DC: Solar Energy Industries Association.

————— 1986c. "Bioenergy Statistics." *Energy Innovation: Development and Status of Renewable Energy Technologies 1985.* Washington, DC: Solar Energy Industries Association.

Federal Energy Regulatory Commission (FERC). 1988. *Hydroelectric Power Resources of the United States.* Washington, DC: GPO.

Finneran, Kevin. 1986a. "Bioenergy: Anaerobic Digestion." *Energy Innovation: Development and Status of Renewable Energy Technologies 1985.* Washington, DC: Solar Energy Industries Association.

————— 1986b. "Bioenergy: Fermentation Ethanol." *Energy Innovation: Development and Status of Renewable Energy Technologies 1985.* Washington, DC: Solar Energy Industries Association.

Fulkerson, William, A.M. Perry, and David B. Reister. 1988. "More Efficient Technologies and Fuel Switching: The Near-Term Prevention Strategy." Energy Division, Oak Ridge National Laboratory. December.

Geller, Howard S. 1989. "National Energy-Efficiency Platform: Description and Potential Impacts." *Energy-Efficiency Issues Paper No. 2.* Washington, DC: American Council for an Energy-Efficient Economy.

General Electric (GE). 1977. "Wind Energy Mission Analysis." Philadelphia: General Electric Space Division. COO/2578-1/1.

Gipe, Paul. 1989. *Wind Energy Comes of Age in California.* Tehachapi, CA: Paul Gipe and Associates.

Goldemberg, Jose, Thomas B. Johansson, Amulya K.N. Reddy, and Robert H. Williams. 1987. *Energy for a Sustainable World.* Washington, DC: World Resources Institute.

Gray, Tom. 1989. "Testimony on S.324, the National Energy Policy Act of 1989." Washington, DC: American Wind Energy Association.

Grubb, Michael. 1988. "The Wind of Change." *New Scientist.* March 17, 1988, p. 43.

Hall, Charles A.S., Cutler J. Cleveland, and Robert Kaufmann. 1986. *Energy and Resource Quality: The Ecology of the Economic Process.* New York: J. Wiley and Sons.

Hansen, Kent, Dietmar Winje, Eric Beckjord, Elias P. Gyftopoulos, Michael Golay, and Richard Lester. 1989. "Making Nuclear Power Work: Lessons from Around the World." *Technology Review.* February/March.

Harats, Y., and D. Kearney. 1989. "Advances In Parabolic Trough Technology In the SEGS Plants." Report presented at 1989 ASME International Solar Energy Conference, San Diego, April 1989. Los Angeles: Luz International Limited.

Harris Poll. 1989. "Favor or Oppose the Building of More Nuclear Power Plants?" Conducted by telephone with 1,248 adults nationwide, 2-6 December 1988. Los Angeles: Creators Syndicate.

Heede, H. Richard, Richard E. Morgan, and Scott Ridley. 1985. *The Hidden Costs of Energy*. Washington, DC: Center for Renewable Resources.

Hohmeyer, Olav. 1988. *Social Costs of Energy Consumption*. Berlin: Springer Verlag.

Houghton, Richard A., and George M. Woodwell. 1989. "Global Climatic Change." *Scientific American*. April.

Hubbard, H.M. 1989a. Solar Energy Research Institute. Testimony before House Committee on Energy and Commerce. April 26.

———— 1989b. "Photovoltaics Today and Tomorrow." *Science*. April 21, p. 297.

International Energy Agency (IEA). 1987. *Renewable Sources of Energy*. Paris: Organization for Economic Co-operation and Development (OECD).

Jensen, C., H. Price, and D. Kearney. 1989. "The SEGS Power Plants: 1988 Performance." Presented at 1989 ASME International Solar Energy Conference, San Diego, April. Los Angeles: Luz International Limited.

Justus, C.G. 1976. "Wind Energy Statistics for Large Arrays of Wind Turbines." Prepared for the National Science Foundation. ERDA/NSF-00547/76/1. Washington, DC. August.

Kendall, Henry W., and Steven J. Nadis (eds.). 1980. *Energy Strategies: Toward a Solar Future*. Cambridge, MA: Ballinger.

Klass, Donald L. 1988. "The U.S. Biofuels Industry." Report presented at the International Renewable Energy Conference. Honolulu. September 18-24. Chicago: Institute of Gas Technology.

Komanoff, Charles. 1988. Personal communication.

———— 1981. *Power Plant Cost Escalation*. New York: Komanoff Energy Associates.

Layton, Patricia. 1989. Oak Ridge National Laboratory. Personal communication.

Lidsky, Lawrence M. 1987. "Safe Nuclear Power and the Coalition Against It." *The New Republic*. December 23.

Lipschutz, Ronnie D. 1980. *Radioactive Waste: Politics, Technology, and Risk.* Cambridge, MA: Ballinger.

Lotker, Michael. 1989. Luz International Limited. Personal communication.

Luz International Limited. 1989. "Luz In Brief." Los Angeles.

Lynd, Lee R. 1989. "Large-Scale Fuel Ethanol from Lignocellulose: Potential, Economics, and Research Priorities." Thayer School of Engineering, Dartmouth College, New Hampshire.

Lynette, Robert. 1989a. "Wind Energy Systems." Speech to the Forum on Renewable Energy and Climate Change. Washington, DC: American Wind Energy Association. June 9.

———— 1989b. Personal communication.

MacKenzie, James J., and Mohamed T. El-Ashry. 1988. *Ill Winds: Airborne Pollution's Toll on Trees and Crops.* Washington, DC: World Resources Institute.

Marland, Gregg. 1989. "Reforestation: Pulling Greenhouse Gases Out of Thin Air." *Biologue.* Vol. 6, No. 2.

Marland, Gregg, and Anthony Turhollow. 1988. "CO_2 Emissions from Production and Combustion of Fuel Ethanol from Corn." Oak Ridge National Laboratory, Oak Ridge, TN.

Marshall Institute. 1989. *Scientific Perspectives on the Greenhouse Problem.* Washington, DC: George C. Marshall Institute.

Marshall, B. W. 1989. Sandia National Laboratory. Personal communication.

Meridian Corporation. 1989. "Energy System Emissions and Materiel Requirements." Report prepared for the Office of Renewable Energy, Department of Energy. February.

Metz, W.D. and A.L. Hammond. 1978. *Solar Energy in America.* Washington, DC: American Association for the Advancement of Science.

Mintzer, I.M. 1987. "A Matter of Degrees: The Potential for Controlling the Greenhouse Effect." *Research Report No. 5.* Washington, DC: World Resources Institute.

Moore, Terry. 1989. "Thin Films: Expanding the Solar Marketplace." *EPRI Journal*, March, p. 4.

National Association of Home Builders (NAHB). 1988. *Solar Electric Houses Today.* Upper Marlboro, MD: NAHB National Research Center.

National Science Foundation (NSF). 1972. *An Assessment of Solar Energy as a National Resource.* NSF/NASA Solar Energy Panel. Washington, DC.

National Wood Energy Association (NWEA). 1989a. "Environmental Regulation of Biomass Facilities." *Biologue.* Vol. 6, No. 1.

—— 1989b. "Federal Programs for Research and Development of Biomass and Municipal Waste Technology." Arlington, VA: NWEA.

—— 1988. "Ethanol Plants." *Biologue.* Vol. 5, No. 2.

Newsday. 1989. *Rush to Burn: Solving America's Garbage Crisis?* Washington, DC: Island Press.

Office of Technology Assessment (OTA). 1985. *New Electric Power Technologies: Problems and Prospects for the 1990's.* OTA-E-246. Washington, DC: US Congress, Office of Technology Assessment. July.

Ogden, Joan M., and Robert H. Williams. 1989. *Solar Hydrogen: Moving Beyond Fossil Fuels.* Washington, DC: World Resources Institute.

Ostlie, David L. 1989. "The Whole Tree Burner: A New Technology In Power Generation." Minneapolis: Energy Performance Systems, Inc.

Pacific Gas and Electric (PG&E). 1988. "Solar Central Receiver Technology Advancement for Electric-Utility Applications." Research and Development Department. San Ramon, CA: PG&E. August.

Paul, Bill. 1989. "U.S. Is Rapidly Losing Its Lead in Alternative Energy." *The Wall Street Journal.* August 15.

Pollard, Robert D. 1987. "Is Nuclear Power Poised to Repeat Past Mistakes?" Cambridge, MA: Union of Concerned Scientists.

Porter, Kevin L. 1989. Investor Responsibility Research Center. Personal communication.

Rader, Nancy. 1989. *Power Surge: The Status and Near-Term Potential of Renewable Energy Technologies.* Washington, DC: Public Citizen. May.

Ramanathan, V. 1988. "The Greenhouse Theory of Climate Change: A Test by an Inadvertent Global Experiment." *Science.* April 15.

Real Goods News. 1989. Catalog of alternative-energy products.

Reid, T.R. 1989. "New Atomic Dump Poses Unprecedented Challenges." *The Washington Post.* July 5.

Rinebolt, David C. 1989a. National Wood Energy Association. Testimony before the House of Representatives Science, Space and Technology Committee. March 2.

———— 1989b. Personal communication.

Robinson, Gordon. 1988. *The Forest and the Trees: A Guide to Excellent Forestry.* Washington, DC: Island Press.

Rogers, Wayne L. 1989. "Hydropower: The Forgotten Technology." Washington, DC: National Hydropower Association.

Rosenfeld, Arthur H. , and David Hafemeister. 1988. "Energy-Efficient Buildings." *Scientific American.* April.

Sandia National Laboratories (SNL). 1987a. *Today's Photovoltaic Systems: An Evaluation of Their Performance.* SAND87-2585. Albuquerque: SNL.

———— 1987b. *The Interconnection Issues of Utility-Intertied Photovoltaic Systems.* SAND87-3146. Albuquerque: SNL.

Schneider, Stephen H., and Rondi Londer. 1984. *The Coevolution of Climate and Life.* San Francisco: Sierra Club Books.

Shea, Cynthia Pollock. 1988. "Renewable Energy: Today's Contribution, Tomorrow's Promise." *Worldwatch Paper No. 81.* Washington, DC: Worldwatch Institute.

Sissine, Fred J. 1989. "Renewable Energy: Federal Programs." CRS Issue Brief. Washington, DC: Congressional Research Service.

Sklar, Scott. 1989. "Programs On Renewable Energy: Solar Contributions In the 1990's and Beyond." Testimony before US Department of Energy National Strategy Hearing. August 1. Washington, DC: Solar Energy Industries Association.

Solar Energy Industries Association (SEIA). 1989. "Federal Programs for Research and Development of Solar Energy Technology: Industry Analysis and Recommendations for Funding for FY 1990." Arlington, VA: SEIA.

———— 1988. "Solar Thermal Energy: Solar Electricity and Heat In The 1990s." Arlington, VA: SEIA.

Sorensen, B. 1976. "Dependability of Wind Energy Generators with Short-Term Energy Storage." *Science*. November 26.

Stommel, Henry, and Elizabeth Stommel. 1979. "The Year Without A Summer." *Scientific American*. June.

Stricharchuk, Gregory. 1989. "Westinghouse, Expecting a Rebound For Nuclear Power, Plans New Plant." *The Wall Street Journal*. April 21.

Toronto World Conference. 1988. *The Changing Atmosphere: Implications for Global Security*. Conference statement.

Turhollow, Anthony. 1989. Herbaceous Energy Crops Program, Oak Ridge National Laboratory. Personal communication.

Victor, Kirk. 1989a. "The Nuclear Turn-On." *National Journal*. September 9.

———— 1989b. "Solar Eclipse." *National Journal*. September 9.

Wald, Matthew L. 1989. "US Companies Losing Interest In Solar Energy." *The New York Times*. March 7.

Williams, Robert H. 1989. "Biomass Gasifier/Gas Turbine Power and the Greenhouse Warming." Princeton, NJ: Center for Energy and Environmental Studies. April.

Williams, Susan, and Kevin Porter. 1989. *Power Plays: Profiles of America's Independent Renewable Electricity Developers*. Washington, DC: Investor Responsibility Research Center.

World Resources Institute (WRI). 1988. *World Resources 1988-89*. New York: Basic Books.

Wright, L.L., J.H. Cushman, and P.A. Layton. 1989. "Dedicated Energy Crops: Expanding the Market Through Improving the Resource." (Draft.) Oak Ridge National Laboratory.

Weisskopf, Michael. 1989. "U.S. Contribution To 'Greenhouse Effect' Rises." *The Washington Post*. September 16.

Yergin, Daniel. 1988. "Energy Security in the 1990's." *Foreign Affairs*. Autumn.